To revise any of these topics more thoroughly, see *Letts Revise GCSE Science Study Guide* (see inside back cover for how to order)

THIS BOOK AND YOUR GCSE EXAMS

Introduction

This book is designed to help you get better results.

- Look at the grade A and C candidates' answers and see if you could have done better.
- Try the exam practice questions and then look at the answers.
- Make sure you understand why the answers given are correct.
- When you feel ready, try the GCSE mock exam papers.

If you perform well on the questions in this book you should do well in the examination. Remember that success in examinations is about hard work, not luck.

What examiners look for

- Examiners are obviously looking for the right answer, however, it does not have to match the wording in the examiner's marking scheme exactly.
- Your answer will be marked correct if the answer is correct Science, even if it is not expressed exactly as it is on the mark scheme. The examiner has to use professional judgement to interpret your answers. You do not get extra marks for writing a lot of irrelevant words.
- You should make sure that your answer is clear, easy to read and concise.
- You must make sure that your diagrams are neatly drawn. You do not need to use a ruler to draw diagrams. Diagrams are often drawn too small for the examiner to see them clearly. They should be clearly labelled with label lines. There are times where a ruler will help e.g. drawing ray diagrams.
- On many papers you will have to draw a graph. At Foundation tier axes of the graph and scales are usually given. However, at Higher tier you usually have to choose them. Make sure you always use over half the grid given and make sure you label the axes clearly. If the graph gives points that lie on a straight line use a ruler and a sharp pencil for this line. If it is a curve, draw a curve rather than joining the points with a series of straight lines. Remember you may have some anomalous points and your line or curve should not go through these.

Exam technique

- You should spend the first few minutes of the examination looking through the whole exam paper.
- Use the mark allocation to guide you on how many points you need to make and how much to write.
- You should aim to use one minute for each mark; thus if a question has 5 marks, it should take you 5 minutes to answer the question.
- Plan your answers; do not write down the first thing that comes into your head. Planning is absolutely necessary in questions requiring continuous or extended answer questions.
- Do not plan to have time left over at the end. If you do, use it usefully. Check you have answered all the questions, check arithmetic and read longer answers to make sure you have not made silly mistakes or missed things out.

DIFFERENT TYPES OF QUESTIONS

Questions on end of course examinations (called terminal examinations) are generally structured questions. Approximately 25% of the total mark, however, has to be awarded for answers requiring extended and continuous answers.

Structured questions

A structured question consists of an introduction, sometimes with a table or a diagram followed by three to six parts, each of which may be further sub-divided. The introduction provides much of the information to be used, and indicates clearly what the question is about.

- Make sure you read and understand the introduction before you tackle the question.
- Keep referring back to the introduction for clues to the answers to the questions.

Remember the examiner only includes the information that is required. For example, if you are not using data from a table in your answer you are probably not answering it correctly.
Structured questions usually start with easy parts and get harder as you go through the question. Also you do not have to get each part right before you tackle the next part.

Some questions involve calculations. Where you attempt a calculation you should always include your working. Then if you make a mistake the examiner might still be able to give you some credit. It is also important to include units with your answer.

Some structured questions require parts to be answered with longer answers. A question requiring continuous writing needs two sentences with linked ideas. A question requiring extended writing may require an answer of six to ten lines. Candidates taking GCSE examinations generally do less well at questions requiring extended and continuous writing. They often fail to include enough relevant scoring points and often get them in the wrong order. Marks are available on the paper here for Quality of Written Communication (QWC).
To score these you need to write in correct sentences, use scientific terminology correctly and sequence the points in your answer correctly.

 This logo in a question shows that a mark is awarded for QWC.

Usually guidance is given about the nature of the written work being assessed, for example, 'One mark is for a clear ordered answer' or 'One mark is for correct spelling, punctuation and grammar' or 'One mark is for correct use of scientific words'.

About 5% of the marks are awarded for questions testing Ideas and Evidence. Usually these questions do not require you to recall knowledge. You have to use the information given to you in the question.

WHAT MAKES AN A/A*, B OR C GRADE CANDIDATE

Obviously, you want to get the highest grade that you possibly can. The way to do this is to make sure that you have a good all-round knowledge and understanding of Science.

GRADE A* ANSWER

The specification identifies what an A, C and F candidate can do in general terms. Examiners have to interpret these criteria when they fix grade boundaries. Boundaries are not a fixed mark every year and there is not a fixed percentage who achieve a grade each year. Boundaries are fixed by looking at candidates' work and comparing the standards with candidates of previous years. If the paper is harder than usual the boundary mark will go down. The A* boundary has no criteria but is fixed initially as the same mark above the A boundary as the B is below it.

GRADE A ANSWER

A grade candidates have a wide knowledge of Science and can apply that knowledge to novel situations. An A grade candidate generally has no bad questions and scores marks throughout. An A grade candidate has to have sat Higher tier papers. The minimum percentage for an A grade candidate is about 80%.

GRADE B ANSWER

B grade candidates will have a reasonable knowledge of the topics identified as Higher tier only. The minimum percentage for a B candidate is exactly halfway between the minimum for A and C (on Higher tier).

GRADE C ANSWER

C grade candidates can get their grade either by taking Higher tier papers or by taking Foundation tier papers. The minimum percentage for a C on Foundation tier is approximately 65% but on Higher tier it is approximately 45%. There are some questions common to both papers and these are aimed at C/D grade level.

If you are likely to get a grade C or D on Higher tier you would be seriously advised to take Foundation tier papers. You will find it easier to get your C on Foundation tier as you will not have to answer the questions targeted at A and B.

HOW TO BOOST YOUR GRADE

Grade booster ---> How to turn C into B

- All marks have the same value. Don't forget the easy marks are just as important as the hard ones. Learn the definitions – these are easy marks in exams and they reward effort and good preparation. If you want to boost your grade, you **cannot** afford to miss out on these marks – they are easier to get.
- Look carefully at the command word at the start of the sentence. Make sure you understand what is required when the word is **state, suggest, describe, explain** etc.
- For numerical calculations, always include units. In Physics there are some equations you must remember and some the examiner can give you.
- If the question asks for the name of a chemical, do not give a formula.
- Read the question twice and underline or highlight key words in the questions, e.g. in terms of covalent bonding, suggest why nitrogen is less reactive than chlorine.
- Use the periodic table and other data that you are given. Don't try to remember data such as relative atomic masses.
- When writing the structural formula of an organic compound make sure that each carbon forms four bonds and you have not missed off hydrogen atoms.

Grade booster ---> How to turn B into A/A*

- Always write balanced equations for chemical reactions. This is one of the ways chemists communicate information. Make sure you use the right symbols and formulae in your equations.
- Make sure that the equation you have written is correctly balanced. Wrongly balanced equations will cost you a mark each time. Ionic equations must have the same total charge on the left-hand side and the right-hand side.
- Check any calculations you have made at least twice, and make sure that your answer is sensible. For example, if you divide 0.49 by 1.9 the answer should be approximately 0.25.
- Make sure you know the difference between number of decimal places and number of significant figures. 25.696 to one decimal place is 25.7 and to 4 significant figures it is 25.70. Make sure you give the correct unit.
- Give complete colour changes. The test for an alkene is NOT that bromine turns colourless, but the colour change is from brown to colourless.
- Try to give conditions for chemical reactions, e.g. concentrated or dilute acid or heat to 140°C.
- In questions requiring extended writing make sure you make enough good points and you don't miss out important points. Read the answer through and correct any spelling, punctuation and grammar mistakes.

CHAPTER 1

Cell structure and division

To revise this topic more thoroughly, see Chapter 1 in *Letts Revise GCSE Science Study Guide*.

Try this sample GCSE question and then compare your answers with the Grade C and Grade A model answers on the next page.

The diagram shows a cell taken from a multi-cellular organism.

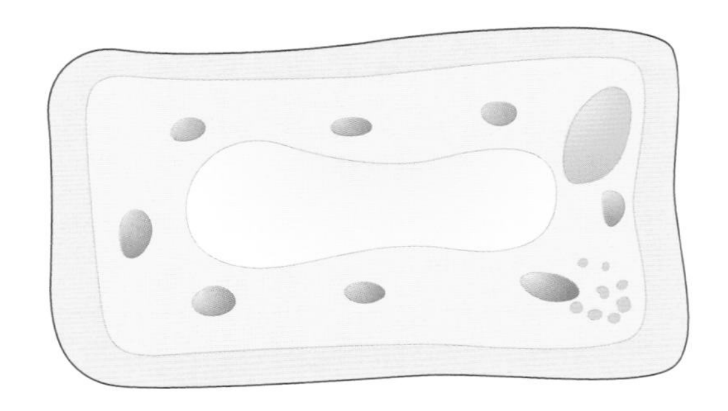

a This cell is from a plant.
Explain how you can tell. **[3]**

b What does selectively permeable mean? **[2]**

c Write down the name of one part of this cell that is selectively permeable. **[1]**

d Desmond put some of these cells into a strong sugar solution.
He then looked at the cell again under the microscope.
This is what he saw.

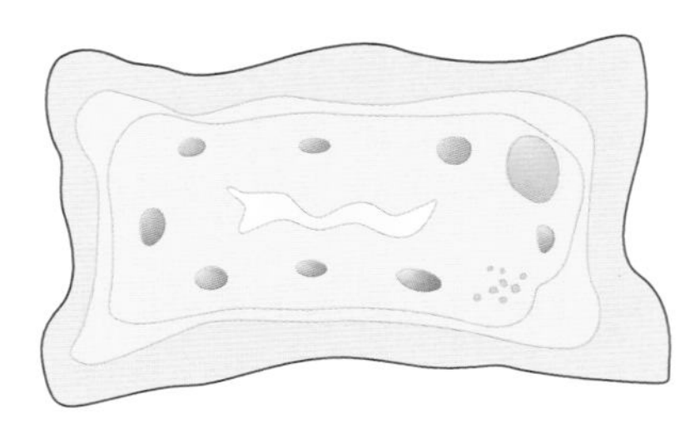

Explain what has happened to the cell. **[4]**

(Total 10 marks)

These two answers are at Grades C and A. Compare which one your answer is closest to and think how you could have improved it.

GRADE C ANSWER

Bishen

a This is because it has a cell wall ✓ and chloroplasts. ✓

Bishen gives two reasons but there are three marks for this so he has missed a reason and gains only two marks.

b This means it lets some substances through but not others. ✓

This is a simple but correct definition – one mark.

c Cell wall ✗

Wrong, the cell wall is permeable – it is the cell membrane that is selectively permeable.

d The cell had shrivelled up because it had lost some of the solution from the vacuole. ✗ This is because the sugar solution is more concentrated than the cell contents ✓ so they had moved out by osmosis. ✓

Bishen realises that the cell has gone flaccid and that this is because of osmosis. He incorrectly thinks that it has lost some of the cell sap, not just water – two marks awarded.

5 marks = Grade C answer

Grade booster ···> move a C to a B

Bishen has a reasonable knowledge of cell structure but he has made a mistake that is common to C grade candidates. When he answers questions about osmosis he must be careful to say that it is the movement of water and not movement of a solution.

GRADE A ANSWER

Venkat

a This is a plant cell because it has a cell wall, chloroplasts and a cell sap vacuole. ✓✓✓

Venkat correctly gives three important structures that are found in plant cells and not in animal cells – three marks.

b A selectively permeable membrane only lets through small soluble molecules and not larger molecules. ✓

This is a better answer than Bishen's but there is only one mark available.

c The cell membrane. ✓

Venkat names the correct structure and gains the mark.

d The cell has lost water ✓ by osmosis. ✓ This is because the cell has a less negative water potential compared to the concentrated sugar solution. ✓

Unlike Bishen, Venkat shows that he understands osmosis and knows that it is water that is moving, not the solution – three marks awarded.

8 marks = Grade A answer

Grade booster ···> move A to A*

Venkat has a good understanding of osmosis and like many A and A* candidates he correctly refers to the water potential of the solutions rather than their concentrations, although this is not necessary in order to score the marks. He fails to give a full explanation of the changes in the cell by not mentioning the loss of turgor pressure.

QUESTION BANK

1 Jack was looking at some pond water and saw a small organism.
He looked in a book and found out that the organism was called hydra.

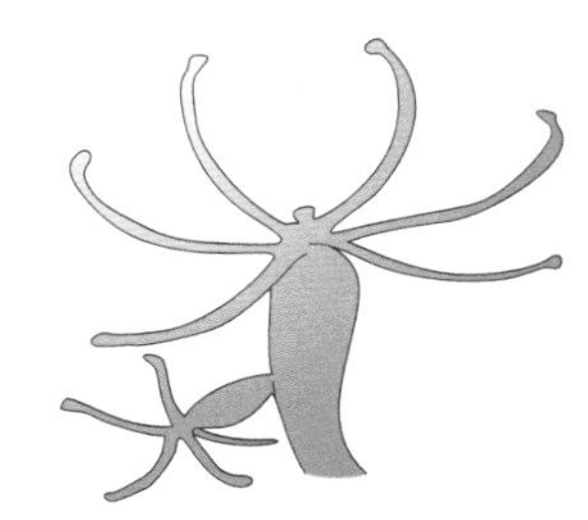

He kept the organism in a tank of water and made a number of observations:

- the hydra moved its tentacles towards small organisms
- the tentacles caught the organisms and placed them into a mouth
- if he shone a bright light at the hydra it would move away
- on the side of the hydra a very similar organism formed.

a) Jack decided that the hydra was showing four of the characteristics of living organisms.
Write down these four characteristics and explain how the hydra is showing each one.

1 ....

2 ....

3 ....

4 .... (4)

b) Jack then used a microscope to find out if the hydra was a plant or an animal.
Explain how he could show that it was an animal.

....

....

.... (3)

c) Jack decided that the small organism that was formed on the side was produced by mitosis.

i) What is mitosis?

....

.... (2)

ii) Where exactly in a cell are genes found?

....

....

.... (2)

TOTAL 11

ANSWERS ON PAGE 14 ANSWERS ON PAGE 14 ANSWERS ON PAGE 14 ANSWERS ON PAGE 14

2 Grace carried out an experiment to demonstrate osmosis.
She used a special material called dialysis tubing to make a bag.
The dialysis tubing is selectively permeable.
She half filled the bag with sugar solution.
Grace then lowered the bag into a beaker of distilled water.

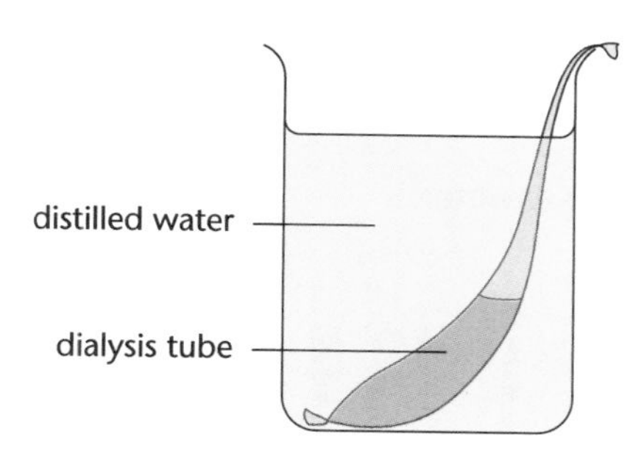

Every five minutes Grace took the bag out of the water and wiped the outside with a tissue.
She then weighed the bag and replaced it in the beaker.
The table shows her results.

Time in minutes	0	5	10	15	20	25	30
Mass in grams	17.0	23.0	29.5	34.0	34.5	35.0	35.0

a) i) Plot the data on the grid provided. ④

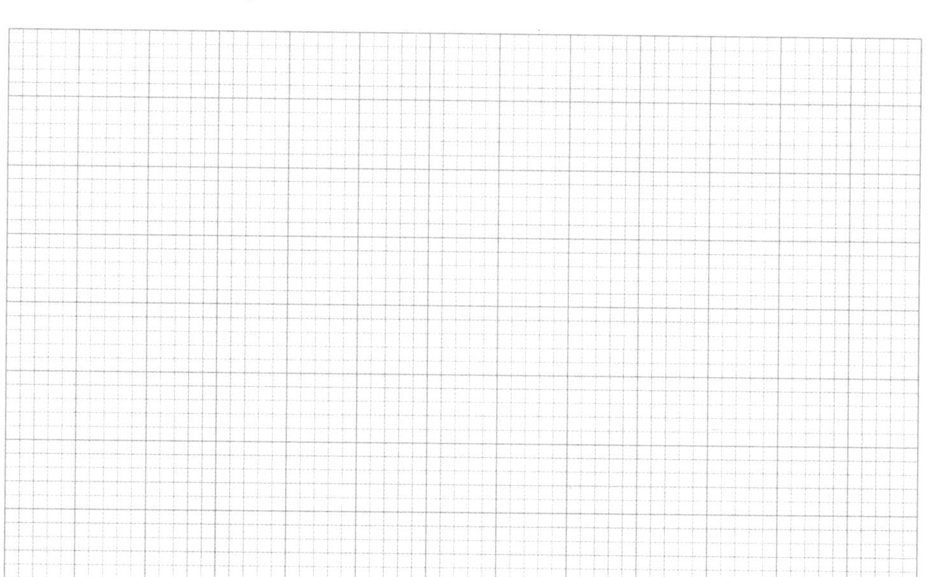

ii) Finish the graph by drawing the best curve through the points. ①

b) Explain why Grace wiped the outside of the bag with tissue paper before she took each result.

....

.... ②

c) Describe what happened to the mass of the bag during the experiment.
Use your graph to help you.

....

....

.... ②

d) Explain why the mass of the bag changed during the experiment.

....

....

....

.... ④

TOTAL 13

---

3 *Valonia* is an alga.
It has large green cells and lives in seawater.

a) Suggest why the cells of *Valonia* are green.

....

.... ②

b) The graph shows the concentrations of mineral ions inside and outside the cells of *Valonia* when it is in seawater.

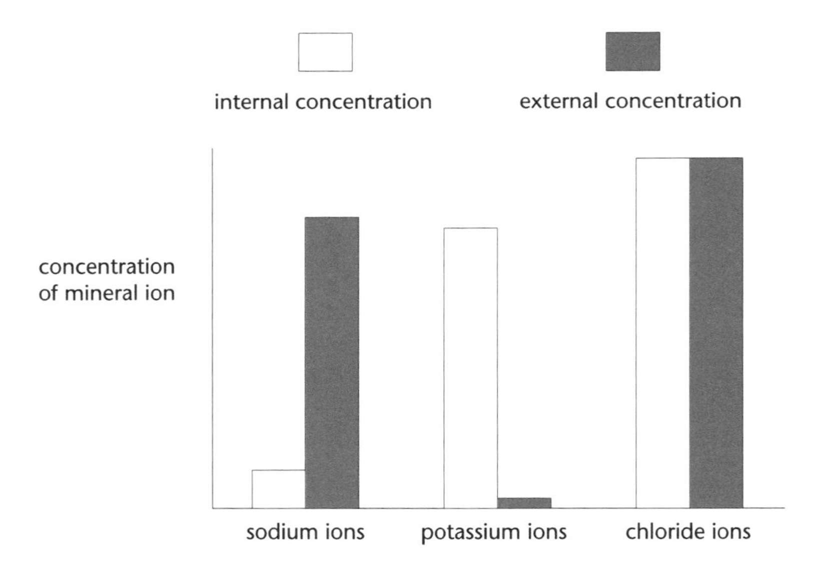

i) *Valonia* has very large cells.
This made it easier to obtain this data.

Suggest why this is.

....

.... ①

ii) Describe how the concentrations of the different mineral ions vary inside, compared to outside the cells.

....

....

....

.... ③

iii) The cells of *Valonia* can pump ions across the cell membrane by active transport. Write down which mineral ions are transported and in which direction.

....

....

.... ②

c) Graph 2 shows the rate at which oxygen is taken up by the *Valonia* cells.
This was measured before and after extra potassium ions were added to the seawater.
One mark is awarded for the correct use of scientific terms.

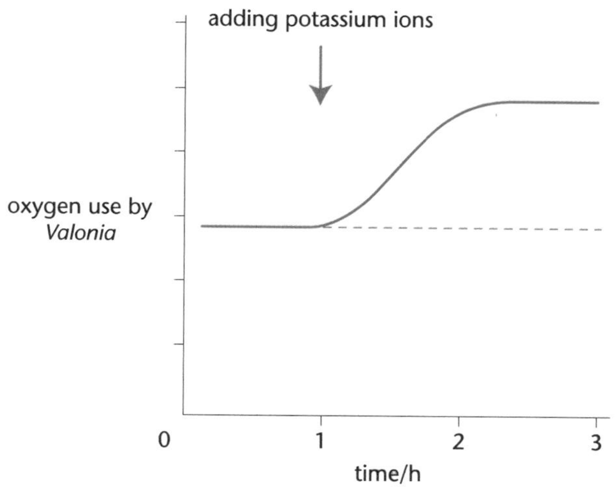

adding potassium ions

oxygen use by *Valonia*

0 1 2 3

time/h

Adding the potassium ions alters the amount of oxygen used by the *Valonia* cells.

Explain why.

....

....

....

.... ④+①

TOTAL 13

ANSWERS ON PAGE 14 ANSWERS ON PAGE 14 ANSWERS ON PAGE 14 ANSWERS ON PAGE 14

QUESTION BANK ANSWERS

1 a) Sensitivity: the hydra is sensitive to light/small organisms (1)
Feeding: the hydra feeds on the small organisms (1)
Movement: the hydra moves away from light/moves its tentacles (1)
Reproduction: small organisms are forming on the side (1)

b) Look at the cells of the hydra under the microscope
If they are animal cells there will be no cell wall
No chloroplasts
No cell sap vacuole any three points (3)

c)i) A type of cell division (1)
That produces genetically identical cells/ Cells with the same number of chromosomes (1)

ii) On the chromosomes (1)
In the nucleus of the cell (1)

Examiner's tip

The question says 'where exactly' and there are two marks for the answer. Many candidates would just write in the nucleus but the answer needs more detail to score full marks.

2 a)i) Mass on the *y*-axis, time on the *x*-axis (1)
Suitable scales (1)
Correct plotting × 2 (2)

Examiner's tip

This question is definitely a higher tier question because you need to decide on which axis to plot time and on which to plot the mass. Remember the variable that Grace measures goes on the y-axis.

ii) Best curve drawn (1)

Examiner's tip

Make sure that you draw a single smooth line. You will lose the mark if you draw a sketchy line or zig-zag from point to point. Remember a best curve may not go through any of the points.

b) To remove any water from the outside of the bag (1)
So that it did not add to the mass (1)

c) Steady increase in mass at first
Mass then becomes constant
Correct use of figures any two (2)

Examiner's tip

It is fairly easy to say that the mass goes up then stays the same but this would only score one mark. To score the other mark you must say that the rise is constant or quote a figure from the graph such as 'the mass starts to level off after 15 minutes'.

d) Bag took in water (1)
By osmosis (1)
Through selectively permeable membrane (1)
Because solution inside bag was more concentrated than the water (1)

3 a) They contain chloroplasts/chlorophyll (1)
In order to photosynthesise/make food (1)

b)i) Makes it easier to take some cytoplasm out of the cell (1)

Examiner's tip

The word 'suggest' in the question means that you are not expected to have learnt the answer but need to think of a reasonable answer, given the facts in the question.

ii) Chloride concentrations the same inside as outside (1)
Sodium more concentrated outside (1)
Potassium more concentrated inside (1)

iii) Sodium ions must be pumped out of the cells (1)
Potassium ions must be pumped in (1)

Examiner's tip

Remember that active transport is used to move substances from a low concentration to a high concentration. Therefore, to build up such a difference in sodium and potassium concentrations, they must have been pumped by active transport.

c) More potassium ions to be pumped into the cell
Pumps have to work faster
Active transport/pumps need energy
Produced by respiration
This requires oxygen any 4 points (4)

The extra mark for quality of written communication is given if you use correct scientific terms and link them together in your answer. (1)

FOR MORE INFORMATION ON THIS TOPIC ... SEE REVISE GCSE SCIENCE ... CHAPTER 1

CHAPTER 2

Humans as organisms

To revise this topic more thoroughly, see Chapter 2 in *Letts Revise GCSE Science Study Guide.*

Try this sample GCSE question and then compare your answers with the Grade C and Grade A model answers on the next page.

It is very important for humans to keep their internal body temperature constant. They have a number of systems that help to keep their temperature constant even if the external temperature changes greatly.

a (i) What is the normal internal body temperature of humans? **[1]**

(ii) Why is it so important for humans to keep their body temperature constant? **[2]**

b The skin contains a number of mechanisms for controlling body temperature.
The two diagrams show sections through the human skin.
One diagram shows the skin when the body temperature is above normal.
The second shows the skin when the body temperature drops below normal.

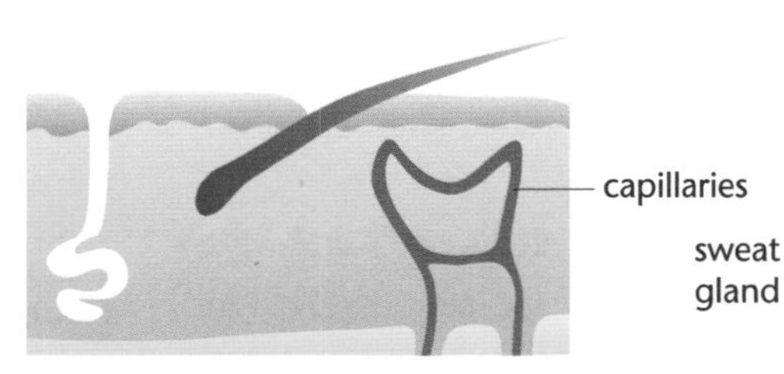

body temperature above normal

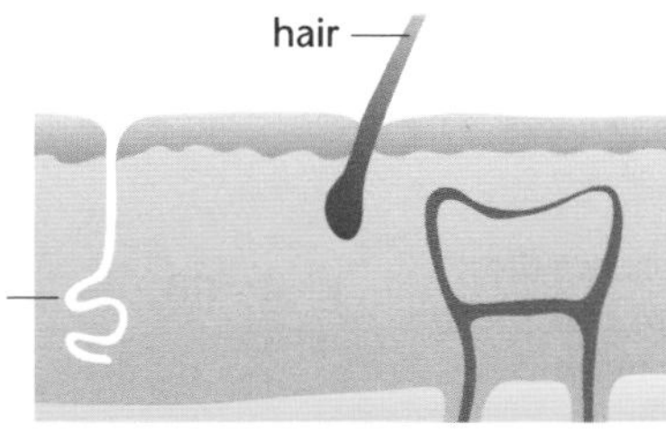

body temperature below normal

(i) Describe what has happened to the blood vessels in the skin when the body temperature has dropped. **[1]**

(ii) Why are the sweat glands important when body temperature goes up? **[3]**

(iii) The hairs in the skin are raised when the body temperature falls.
This has little effect in humans but can help other mammals.
Explain how. **[3]**

(Total 10 marks)

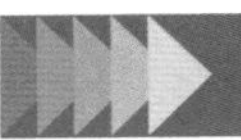

These two answers are at Grades C and A. Compare which one your answer is closest to and think of how you could have improved it.

GRADE C ANSWER

Nnena

a (i) 37°C ✓

(ii) If our temperature goes too high it will kill our enzymes ✗

b (i) The capillaries have moved away from the skin. ✗

(ii) The sweat glands produce more sweat. ✓ This helps to take more heat from the skin letting us cool down. ✓

(iii) The hairs trap more air when they stand up. ✓ This helps to keep the hairy animal warm. ✗

Nnena knows that normal body temperature is 37°C – one mark.

Nnena knows that the answer has something to do with enzymes but enzymes cannot be killed.

This is a common mistake. Many candidates think that blood vessels can move.

The important point that Nnena misses is that the sweat evaporates. This removes heat. She gains two marks out of three.

Again, Nnena knows that the hairs trap air but does not link it to being a good insulator. She says that it keeps the animal warm but does not say that it reduces heat loss. Only one mark.

4 marks = Grade C answer

Grade booster ····> move a C to a B

Nnena has a reasonable understanding of temperature regulation but made several very common errors for a C/D candidate. Enzymes are protein molecules so cannot be killed and blood vessels in the skin do not move, they just open or close.

GRADE A ANSWER

Amber

a (i) 37°C ✓

(ii) If the body temperature goes too high it will effect our enzymes. ✓

b (i) The capillaries closed down. This is called vaso-constriction. ✓

(ii) The sweat glands make more sweat. ✓ This evaporates from the skin taking heat with it ✓ and so cooling us down. ✓

(iii) In hairy mammals the hair stands up and traps more air. ✓ This insulates the animal making it lose less heat. ✓✓

Again the temperature is correct – one mark.

Amber knows that enzymes are affected by temperature but does not say how – only one mark out of two.

Amber understands the process of vaso-constriction and gains the mark.

Again Amber scores full marks because she appreciates the importance of evaporation.

Full marks again. Amber knows that air is a good insulator reducing heat loss.

9 marks = Grade A answer

Grade booster ····> move A to A*

This is certainly an A grade answer and if this standard is repeated throughout the paper, Amber could achieve an A* grade. The only mark she lost was in describing the action of heat on enzymes. She was not precise enough as she did not say that high temperatures prevent the enzymes from working or denature them.

QUESTION BANK

1 Read the following passage carefully and then answer the questions that follow.

> From about 1900 to 1947 the number of people that were recorded as having lung cancer was increasing. Many people put this down to improved scientific understanding or the increasing number of elderly people in the population. Two scientists, Richard Doll and Austin Bradford Hill, decided to investigate this data.
>
> Doll was particularly interested in statistics and so designed a study to look at over 3500 patients. These patients had all been admitted to hospital with some form of cancer. Hill and Doll found that smoking had little effect on most cancers but increased the risk of lung cancer significantly.
>
> They followed this up with a study on 40 000 doctors. They studied them over five years. They found that the risks of suffering from lung cancer increased in roughly the same proportion to the number of cigarettes smoked.

a) Suggest why 'improved scientific understanding' meant that more people were being recorded as having lung cancer.

.... (1)

b) Why was the number of elderly people in the population increasing?

....

.... (1)

c) How did Doll's interest in statistics affect the way that he designed his investigations?

....

.... (1)

d) As well as lung cancer, smoking has been shown to increase the chance of suffering from various other diseases.
Write down the names of **two** of these diseases and explain how they harm the body.

Disease 1 ....

....

....

Disease 2 ....

....

.... (4)

TOTAL 7

---

2 Until 1628, people thought that the blood was pumped by the heart.
They thought that the blood carried oxygen to the tissues and then passed back to the heart in the same vessels.

a) Explain what was correct and what was incorrect with these ideas.

....

....

.... (3)

ANSWERS ON PAGE 21 ANSWERS ON PAGE 21 ANSWERS ON PAGE 21 ANSWERS ON PAGE 21

In 1628 William Harvey published the results of a series of experiments.
He tied a cord around the top of his arm.

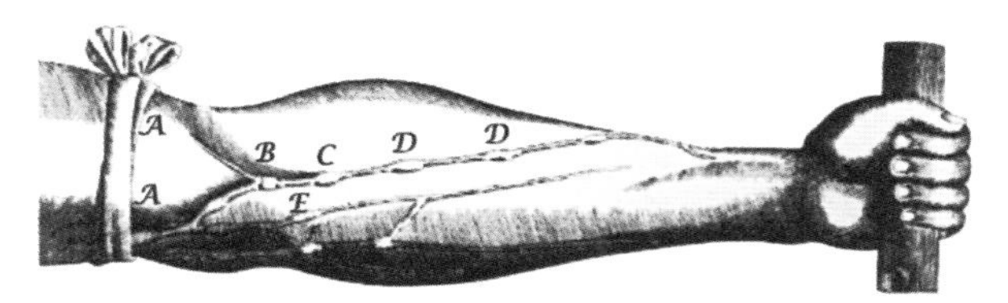

b) He found that the veins below the cord became swollen with blood.
Explain why this is.

....

.... (1)

Harvey then emptied blood from part of the vein by rubbing his finger back in the direction O to H on his drawing.

He found that blood in the part G to O was stopped from flowing backwards by small structures in the vein.

c) i) What are these structures called?

.... (1)

ii) Explain how they stop the blood in the vein from flowing backwards.

....

....

.... (2)

d) Harvey also predicted that small vessels must exist that join arteries to veins.

i) What are these vessels called?

.... (1)

ii) Harvey was unable to prove that these vessels were there.
They were discovered years later by an Italian scientist called Malpighi.
Suggest why Malpighi was able to show that the vessels were there whilst Harvey could not.

....

....

.... (2)

TOTAL 10

 ANSWERS ON PAGE 21 ANSWERS ON PAGE 21 ANSWERS ON PAGE 21 ANSWERS ON PAGE 21

3 The diagram shows some of the main hormone producing glands of the body.

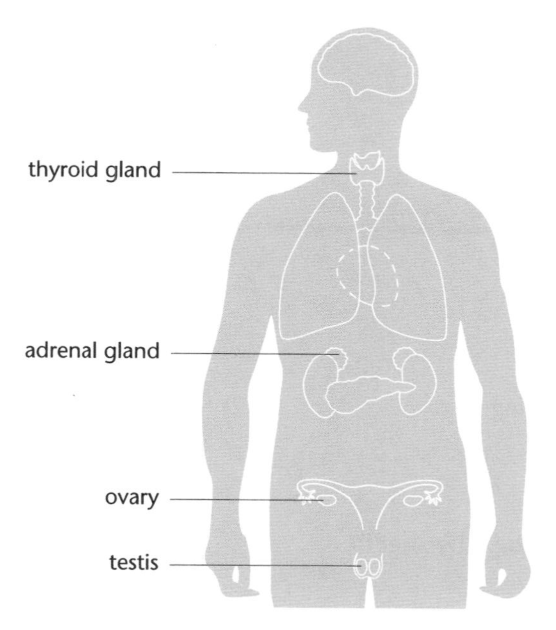

a) Some of the glands have not been labelled.
Add labels to the diagram to show the position of:

i) the pancreas

ii) the pituitary gland. ②

b) The table shows some of the hormones that are produced by these glands.
It also shows some of the functions of these hormones.
Finish the table by writing in the six blank boxes.

Gland	Hormone	Function
	ADH	Controls the concentration of the blood
	Insulin	
Ovary		Repairs the wall of the uterus
	Testosterone	

⑥

c) The table states that the hormone ADH controls the concentration of the blood.
Explain how this works if the blood becomes too concentrated.

....

....

....

....

.... ④+①

TOTAL 13

4 The diagram shows a type of nerve cell (neurone).

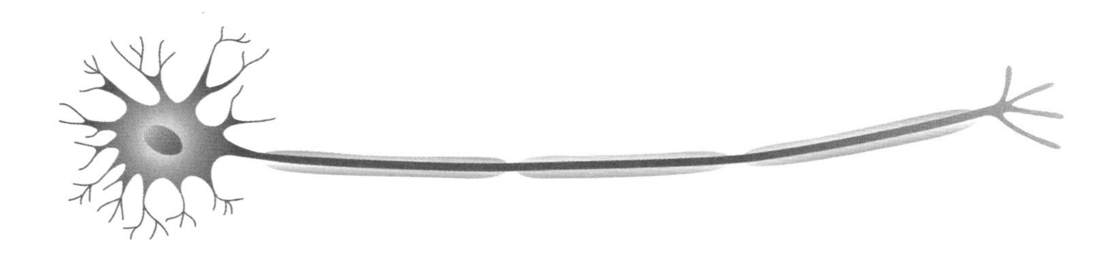

a) Finish the diagram by labelling:

i) the fatty (myelin) sheath

ii) the cell body. ②

b) Write down the function of the fatty sheath.

....

.... ①

c) Write the letter **A** on the part of the neurone that passes information on to a muscle. ①

d) Some people have a disease called motor neurone disease.
Their motor neurones are gradually destroyed.
These people find it difficult to move but can feel, smell and see perfectly normally.
They often find that it becomes difficult to breathe.

i) Explain why people have motor neurone disease can feel, smell and see normally.

....

....

.... ②

ii) Explain why these people find breathing difficult.

....

....

.... ②

TOTAL 8

ANSWERS ON PAGE 21 ANSWERS ON PAGE 21 ANSWERS ON PAGE 21 ANSWERS ON PAGE 21

QUESTION BANK ANSWERS

1 a) The cause of death could be more accurately determined 1

Examiner's tip

This ideas and evidence question is about different ways of interpreting scientific data.

b) There were improvements in medical treatment
Improvements in sanitation/housing/diet any 1 1

c) He studied large numbers of patients/ Followed them over a long period of time 1

d) Bronchitis: 1
excess mucus production/cough 1
Emphysema: 1
destruction of the walls of alveoli/reduced gaseous exchange 1
Heart disease: 1
heart attacks 1

2 a) Correct:
Heart pumps blood 1
Carries oxygen to the tissues 1
Incorrect:
Does not passes back to the heart in the same vessels 1

Examiner's tip

The question asks for both correct points and incorrect points, make sure that you include both.

b) Blood in the veins is passing back to the heart 1

c)i) Valves 1

ii) Flaps close to stop the blood flowing backwards 1
Forced to close by the pressure of the blood 1

Examiner's tip

This is a very common question. Weaker candidates would just say that the flaps close. To get full marks you must say that it is the pressure of the blood trying to flow backwards that closes the flaps.

d)i) Capillaries 1

ii) Need microscopes to see capillaries 1
Microscopes had been improved 1

Examiner's tip

This ideas and evidence question is about how scientists' work is affected by when it took place.

3 a) Correct positioning of labels 2

Examiner's tip

On higher tier papers you must be able to draw in the label lines yourself. They will not be given for you.

b) Pituitary gland
Pancreas
Controls blood sugar level
Oestrogen
Testis
Stimulates the male secondary sexual characteristics 6

c) Concentration of the blood detected in the brain/hypothalamus
More ADH produced
Passes to kidney in the blood
Increases permeability of tubule to water
More water reabsorbed back into the blood
Urine becomes more concentrated/lower in volume any 4 4
The extra mark for quality of written communication is given if you use correct scientific terms and link them together in your answer. 1

Examiner's tip

Many candidates get this the wrong way round. They say that ADH is produced if the blood gets less concentrated. Remember that ADH stands for ANTI diuretic hormone. This means it makes you pass less urine.

4 a) Correct labelling on the diagram 2

b) Insulates the neurone/speeds up the impulse 1

c) **A** on position of motor end plate 1

d)i) Sensory information carried via sensory neurones 1
These are not destroyed by the disease 1

Examiner's tip

Remember to use the word neurone or nerve cell. Many candidates use the word nerve incorrectly. A nerve may contain thousands of neurones, some of which may be sensory and some may be motor.

ii) Intercostal muscles/diaphragm contract during breathing 1
Need impulses from motor neurones in order to contract 1

CHAPTER 3

Green plants as organisms

To revise this topic more thoroughly, see Chapter 3 in *Letts Revise GCSE Science Study Guide.*

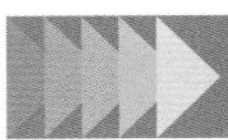

Try this sample GCSE question and then compare your answers with the Grade C and Grade A model answers on the next page.

Rowena found some pond-weed in her garden pond.
She put it in a test tube of water.
She noticed that when a light was shone at the test tube, the pond-weed released bubbles.

a Rowena decided that the pond-weed must be releasing oxygen and that this was produced in photosynthesis.

Finish the balanced equation for this process.

.... + ... → .... + $6O_2$ **[2]**

b Rowena then dropped several ice cubes into the water.

After a while she found that the rate at which the bubbles were released slowed down.

Explain why. **[2]**

c Rowena then decided to investigate the effect of making the light brighter.

She gradually made the light brighter and observed the rate at which the bubbles were given off.

(i) Describe what she would see. **[2]**

(ii) Explain why this happened. **[3]**

(Total 9 marks)

These two answers are at Grades C and A. Compare which one your answer is closest to and think how you could have improved it.

GRADE C ANSWER

Grenville knows the chemicals involved but cannot balance the equation – only one mark.

Grenville does not realise that there is a limit to how fast the pond weed can photosynthesise. One mark out of two.

Grenville

a $H_2O + CO_2 \rightarrow C_6H_{12}O_6 + 6O_2$ ✓ ✗

b The pond-weed photosynthesised slower because the water was colder. ✓

c (i) The pond weed carried on bubbling but the bubbles came off faster as the light got brighter. ✓

(ii) As the light became brighter it provided the plant with more energy. ✓ The extra energy let the pond weed photosynthesise faster. ✓

He scores a mark for knowing that plants photosynthesise more slowly in cold water, but does not say why.

Two out of the three marks are scored for knowing that the brighter light provides more energy for photosynthesis.

5 marks = Grade C answer

Grade booster ····> move a C to a B

Grenville must be able to write out the balanced symbol equation for photosynthesis (and respiration) in order to improve his grade. One of these two equations is often asked and one is the reverse of the other.

GRADE A ANSWER

A correct, balanced equation – full marks.

Again this is a better answer because Sue realises that there is a limit to how fast the weed can bubble.

Sue

a $6H_2O + 6CO_2 \rightarrow C_6H_{12}O_6 + 6O_2$ ✓✓

b The ice makes the temperature of the water drop. This makes photosynthesis go slower. ✓ This is because photosynthesis needs enzymes and they work slower when it is colder. ✓

c (i) As the light got brighter the pond-weed bubbled faster. ✓ When the light got very bright making it brighter still made no difference. ✓

(ii) Brighter light provides the pond-weed with more energy for photosynthesis ✓ so it can photosynthesise faster. ✓ When it gets too bright it damages the plant. ✗

This is a better answer than Grenville's because Sue explains exactly what the drop in temperature does – full marks.

Sue cannot explain why the rate of photosynthesis is limited. In fact, if the pond weed was damaged as she said, the rate would drop. She gains two marks out of three.

8 marks = Grade A answer

Grade booster ····> move A to A*

Sue can write the balanced equation for photosynthesis and knows that if you increase one of the factors needed, the rate will increase up to a certain point. In order to become an A* student she needs to be able to explain why it levels off. She must learn about limiting factors.

QUESTION BANK

1 An experiment was performed to investigate how shoots respond to light.
The tip was removed from a growing shoot.
A jelly block containing a certain concentration of auxin was placed on the cut end of the shoot as shown.
After a period of time the angle of curvature of the shoot was measured.
The process was repeated for different concentrations of auxin.

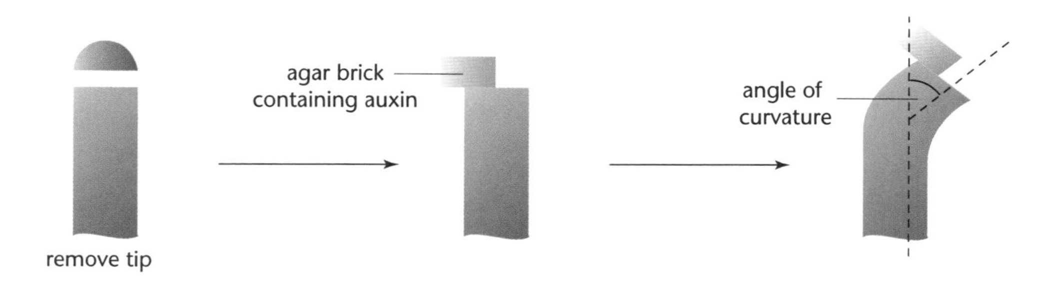

Auxin concentration mg/dm³	0.05	0.1	0.15	0.2	0.25	0.30
Angle of curvature	4	9	13	19	22	17

a) i) Choose a suitable scale for the angle of curvature on the axes shown. (1)

ii) Plot the results of the experiment on the graph. (2)

iii) Finish the graph by drawing the best curve. (1)

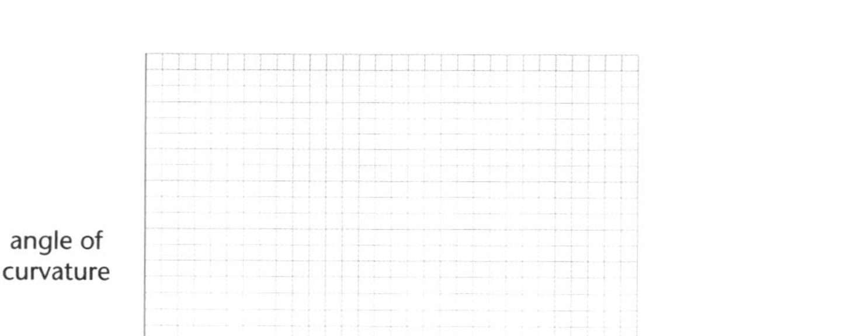

ANSWERS ON PAGE 28 ANSWERS ON PAGE 28 ANSWERS ON PAGE 28 ANSWERS ON PAGE 28

iv) The auxin in the agar block causes the stem to curve over.
What effect does the concentration of auxin have on the amount of bending?

....

....

.... ③

b) i) How does adding auxin to one side of the shoot cause it to curve over?

....

.... ①

ii) What stimulus in nature might cause this type of response in a shoot?

.... ①

iii) How does the plant gain from this type of response?

....

.... ②

c) Gardeners often use plant growth substances, such as auxins.
Describe one way in which they use them.

....

....

.... ③

TOTAL 14

---

2 In the early seventeenth century a Dutch scientist called van Helmont carried out a famous experiment.
At that time people thought that plants obtained their food from the soil.
He grew a tree in a pot of soil, supplying it only with rain water.

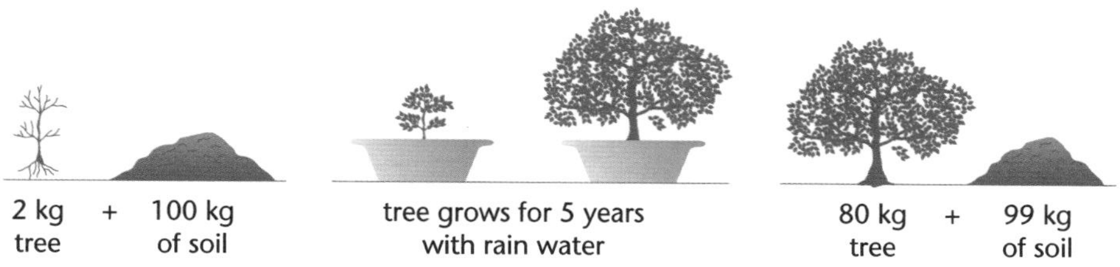

2 kg tree + 100 kg of soil

tree grows for 5 years with rain water

80 kg tree + 99 kg of soil

a) i) Explain how van Helmont disproved the idea that plants gain their food entirely from the soil.

....

.... ②

ii) Van Helmont concluded that the tree gained in mass entirely from the rain water. To what extent is this true?

....

....

.... ③

c) In 1953 an American scientist called Melvin Calvin worked out many of the reactions that enable a plant to increase in mass.
Describe how he would communicate his ideas to other scientists.
How would these methods differ from those used by van Helmont?

....

....

....

.... ③

TOTAL 8

---

3 Rosie notices that if leaves are removed from a plant they dry out and shrivel up.
She decides to design an experiment to investigate transpiration.

a) Explain what is meant by transpiration.

....

.... ②

She picks three leaves from a plant.
She coats one with nail varnish on the top surface of the leaf.
On the second she paints the bottom surface.
The third leaf she leaves unpainted.
Rosie then uses a very accurate balance to measure the mass of the leaves.
She then hangs the leaves from a piece of string.

Rosie re-weighs the leaves at regular intervals.
Her results are shown on the graph.

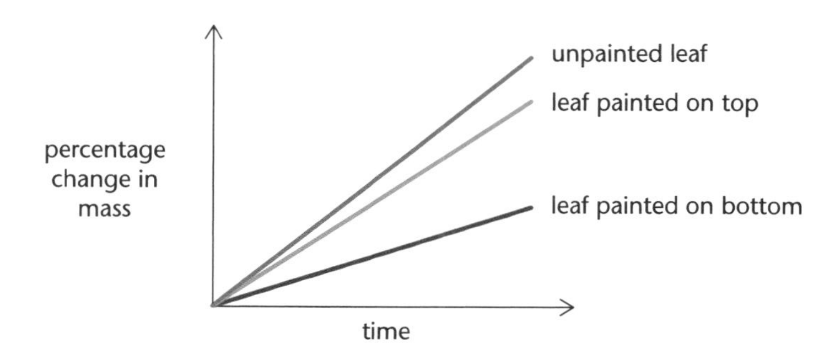

ANSWERS ON PAGE 28 ANSWERS ON PAGE 28 ANSWERS ON PAGE 28 ANSWERS ON PAGE 28

b) i) What happens to the mass of the leaf with no nail varnish?

....

.... ②

ii) Why does the mass of this leaf change?

....

.... ②

iii) The other two leaves change mass at different rates.
Explain why. Add one mark for a clearly ordered answer.

....

....

....

.... ④+①

c) The results of Rosie's experiment would have been different if she pointed a fan at the leaves.

Explain why.

....

.... ②

TOTAL 13

ANSWERS ON PAGE 28 ANSWERS ON PAGE 28 ANSWERS ON PAGE 28 ANSWERS ON PAGE 28

QUESTION BANK ANSWERS

1 a)i) suitable scale chosen 1

Examiner's tip

In this graph you are told which variable to plot on each axis but have to decide on one of the scales. Make sure that you choose a scale that uses at least half of the graph paper. This is what the examiner will be looking for.

ii) Correct plots ×2 2
iii) Best curve drawn 1
iv) Increased concentration makes the shoot bend more 1
Up to a certain concentration after which it stays constant/drops slightly 1
Quote maximum angle of curvature or optimum auxin concentration 1

Examiner's tip

Again, notice that there is a mark for quoting a figure from the graph.

b)i) Makes it grow faster/makes the cells elongate more 1
ii) Light shining from one side 1

Examiner's tip

Just the word 'light' here is not specific enough. It has to be unevenly shining from one side.

iii) Can grow towards the light/trap more sunlight 1
More photosynthesis 1
c) Cut length from stem 1
Dip stem into hormone rooting powder 1
Stem is planted and grows roots 1
or
Included in weedkillers 1
Cause rapid growth 1
Plant uses up food reserves and dies 1
either method for 3 marks

Examiner's tip

There are other uses such as producing seedless fruits or ripening fruits, but these two are the most common ways that gardeners use plant hormones.

2 a)i) The soil lost less mass than the tree gained 1
Correct use of figures 1
ii) Photosynthesis does use water 1
But also uses carbon dioxide 1
Plant also uses some minerals from the soil 1

Examiner's tip

The way the question is worded gives you a clue that there are true parts to the statement but also false parts.

c) He would use:
Telephone
Scientific publications 2
van Helmont would have to use letters/word of mouth 1

Examiner's tip

This ideas and evidence question is about the way in which scientists communicate their ideas and discoveries.

3 a) Loss of water
By evaporation
Mainly from the leaves of plants any 2 2

Examiner's tip

This is an important definition. Make sure that you learn it!

b)i) The leaf loses mass 1
At a constant rate 1

Examiner's tip

Again it is easy to score the first mark, but to score full marks you must say more than just losing mass.

ii) The leaf is losing water 1
By transpiration 1
iii) Both leaves that were painted lost mass more slowly 1
Because the nail varnish blocked transpiration 1
The leaf painted on the bottom lost least 1
Because most stomata are on the bottom of the leaf 1
Plus one for a clearly ordered answer 1
c) The leaves would have lost mass faster
Transpiration rate is faster in the wind
Wind blows away the water vapour
any two 2

FOR MORE INFORMATION ON THIS TOPIC ... SEE REVISE GCSE SCIENCE ... CHAPTER 3

CHAPTER 4

Variation, inheritance and evolution

To revise this topic more thoroughly, see Chapter 4 in *Letts Revise GCSE Science Study Guide*.

Try this sample GCSE question and then compare your answers with the Grade C and Grade A model answers on the next page.

Gregor Mendel was a monk who lived in the nineteenth century.
He worked in a small monastery in Austria.
He carried out various breeding experiments on pea plants.
He discovered that the height of pea plants was controlled by a single gene.
The allele T made the plants tall and the allele t made them short.

a Mendel crossed a homozygous tall pea plant with a dwarf plant.
What would be the genotype of the pea plants that were produced?
Use a genetic diagram to help you. **[3]**

b Before Mendel's time most scientists thought that plants produced by sexual reproduction looked like a mixture of the two parents.
How did Mendel's experiment disprove this? **[2]**

c Mendel's results remained undiscovered or unbelieved for about fifty years.
Suggest why it took so long for his results to be accepted. **[2]**

d A scientist tried to repeat Mendel's experiments.
He performed the same cross and planted the seeds that were produced in several different areas.
He found that the plants grew to slightly different heights in the different areas.
He decided that Mendel's conclusions must have been wrong.
Explain why this scientist was wrong and Mendel was correct. **[2]**

(Total 9 marks)

These two answers are at Grade C and A. Compare which one your answer is closest to and think how you could have improved it.

GRADE C ANSWER

Edmund has carried out the cross correctly and knows that the gametes have only one allele. He has, however, given the phenotype instead of the genotype of the offspring. He gains two marks out of three.

A correct point but there are many other points. These two points are too similar for two marks.

Edmund

a parents TT x tt ✓
gametes T t ✓
genotype tall ✗

b Mendel's plants were all tall. ✓

c Perhaps he did not write them down or publish them in a book. ✓

d Mendel was much more careful when he measured the height of the plants. The scientist made some mistakes. ✗

He correctly says that the plants are tall but what would the other theory predict?

Edmund does not realise that the height of the plants is also influenced by the environment, not just the genes – one mark.

4 marks = Grade C answer

Grade booster ···> move a C to a B

Edmund can carry out genetic crosses but he needs to learn the genetic definitions. He also needs to read the question carefully and make sure that he has given enough separate points in his answer in order to score all the marks.

GRADE A ANSWER

Trevor understands the difference between an organism's genotype and phenotype – three marks.

Trevor gives a different reason than Edmund, but still only gives one reason and gains only one mark.

Trevor

a parents TT x tt ✓
gametes T t ✓
genotype Tt ✓

b If the old theory was right the peas should have been half-way between tall and short. ✓
They were all tall so the old theory was wrong. ✓

c Mendel only worked in a small monastery and was not very well known. ✓

d Perhaps the scientists grew the plants in areas which had different amounts of sunlight. ✓

A good answer. Trevor explains how the fact that all the plants were tall disproves the old theory – full marks.

Trevor realises that the plants may have been grown under different conditions and gets some credit for this – one mark.

7 marks = Grade A answer

Grade booster ···> move A to A*

Trevor has a good understanding of genetics, which is often considered a difficult topic. In the last part of the question he does not explain that an organism's phenotype is a result of its genes and the environment. He simply gives an example.

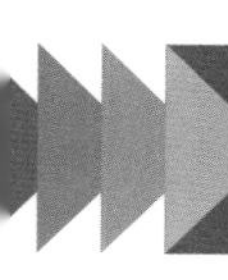

QUESTION BANK

1 Read the following passage carefully and use it to help you to answer the questions that follow.

> Albinism is a condition that causes a person to produce very little coloured pigment in the skin, hair or iris. This means that they have white hair, pink irises and a very pale skin.
>
> The condition is not usually life threatening if the person takes some simple precautions.
>
> Albinism causes the person to become sun burnt very easily and the action of ultra-violet light on the skin is more likely to cause mutations in the cells.
>
> The condition is caused by a recessive allele (a). The dominant allele (A) causes normal pigment production.

a) The action of ultra-violet light on the skin may causes mutations.
What are mutations?

....

.... (1)

b) Suggest **one** 'simple precaution' that a person with albinism might take.

.... (1)

c) The allele that causes albinism is said to be recessive.
What does this mean?

....

....

.... (2)

d) What is the genotype of a person who has albinism?

.... (1)

e) Explain how two parents who do not have albinism can produce a child that does have albinism.
Use a genetic diagram to help you.

(4)

f) Another rare skin condition is called xeroderma.
People that have this condition have freckles that are very sensitive to the sun.
This condition is caused by a different allele.
All children that develop this condition have at least one parent who has the condition.
Suggest and explain why this is.

....

....

.... (2)

TOTAL 11

2 Scientists believe that man evolved from ape-like animals several million years ago.
One important development was the necessary bone structure to walk upright on two feet.

Scientists think that several millions of years ago the earth became drier and forests were replaced by grasslands.
Before this time all apes walked on four feet.
In the grassland, populations of apes appeared that walked on two feet and they increased in numbers.
It is thought that being able to stand upright in grass enabled the ape to see further.

a) Suggest why it might be an advantage for the upright ape to see further.

.... (1)

b) Explain how the population of apes that lived in the grassland changed to walk upright.
Use your knowledge of natural selection and the information above to help you.
One mark is awarded for the correct use of scientific terms.

....

....

....

....

.... (4)+(1)

c) The idea of natural selection was put forward by Charles Darwin.
Before his theory became accepted, most people believed that God created the Earth about 6000 years ago.

i) Darwin's theory needed the Earth to be millions of years old.
Why is this?

....

....

.... (2)

ANSWERS ON PAGE 35 ANSWERS ON PAGE 35 ANSWERS ON PAGE 35 ANSWERS ON PAGE 35

ii) When Darwin published his theories, they caused much controversy.
Explain why this was.

....

....

.... ②

TOTAL 10

---

3 The diagram shows a type of plant called a spider plant.
One plant can reproduce on its own by growing new plants at the end of shoots.

a) What type of reproduction is this plant showing?

.... ①

b) The new plant is produced by cells dividing.
What is the name given to this type of cell division?

.... ①

c) The plants produced by this type of reproduction are often called clones.
What is a clone?

....

.... ②

d) Write down two advantages to a gardener of producing new plants by this type of reproduction rather than by planting seeds.

1. ....

2. .... ②

TOTAL 6

---

4 The diagram shows the chromosomes from one cell of an adult man.

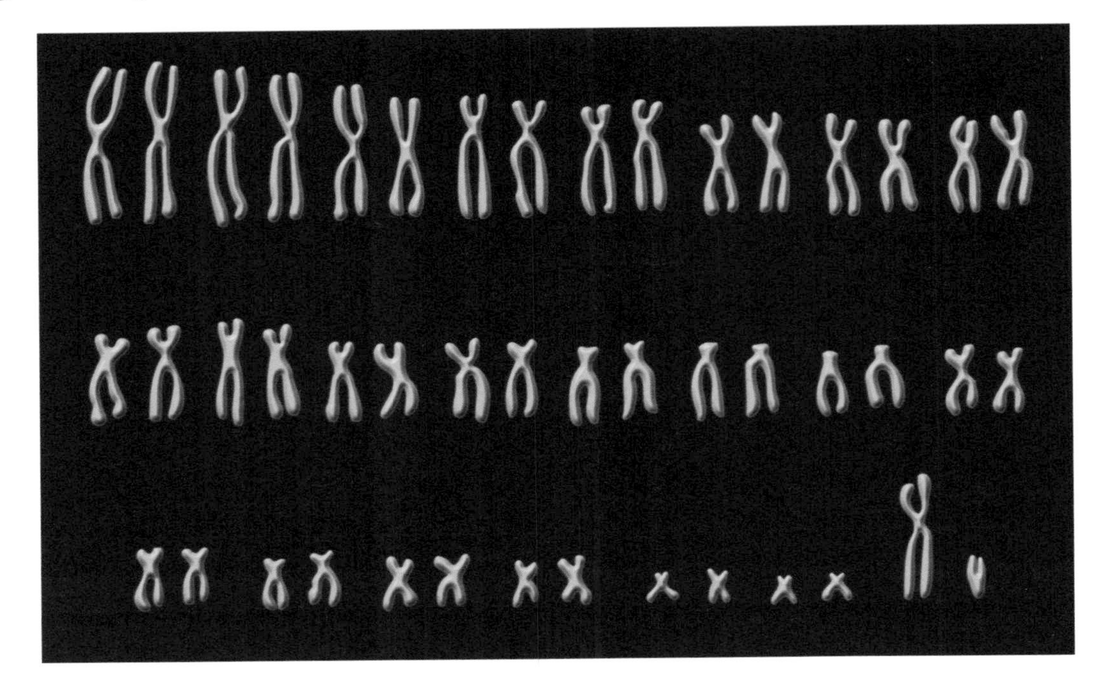

a) How can you tell that the cell comes from a male?

....

.... (2)

b) The chromosomes are from a body cell.
How would they differ if they were from a gamete?

....

....

.... (2)

c) In the testes of males, body cells can divide to produce gametes.

i) Write down the name of this type of cell division.

.... (1)

ii) Sometimes this cell division produces errors in the chromosomes.
Write down the name used to describe these errors.

.... (1)

d) Scientists believe that it may be possible to treat such diseases by genetic engineering.

i) What is genetic engineering?

....

....

.... (2)

ii) Some people think that genetic engineering should not be used on humans.
Suggest why they believe this.

....

....

....

.... (3)

TOTAL 11

ANSWERS ON PAGE 35 ANSWERS ON PAGE 35 ANSWERS ON PAGE 35 ANSWERS ON PAGE 35

QUESTION BANK ANSWERS

1 a) Change in the genetic material/DNA **1**
b) Use suntan lotion/wear sunglasses **1**
c) Does not express itself in the offspring **1**
When together with a dominant allele/ unless two recessives are together **1**
d) aa **1**

Examiner's tip

Often the examiner will tell you which letters to use. If so, then use them! If not, then use capital and small case versions of the same letter. Try and choose a letter where the capital and small case look different so they cannot be confused (such as A and a).

e)

parents	Aa	×	Aa	**1**
gametes	A or a	×	A or a	**1**
offspring	AA	Aa	Aa aa	**1**

aa is an albino **1**

f) Must be caused by a dominant allele **1**
Cannot have the allele without having the disease **1**

Examiner's tip

The best candidates know when to use the word allele and when to use the word gene. An allele is a particular form of a gene, so albinism and xeroderma are caused by particular alleles.

2 a) See predators/prey easier **1**
b) All apes born show variation/some more upright than others
Due to mutations
More upright more likely to survive/survival of the fittest
Reproduce and pass on their genes
Next generation will be more upright
Process continues for many generations
any four **4**
The extra mark for quality of written communication is given if you use correct scientific terms and link them together in your answer. **1**

Examiner's tip

These natural selection questions all have a similar mark scheme. Just apply this to the particular example that you are given in the question.

c) i) Relies on natural selection **1**
This involves gradual changes **1**
ii) Many people were deeply religious
Taken as an alternative to the biblical explanation
People thought that the theory implied that man was descended from monkeys
any two **2**

Examiner's tip

This ideas and evidence question points out that the way in which people are brought up will alter how they accept new scientific ideas.

3 a) Asexual reproduction/vegetative reproduction **1**
b) Mitosis **1**

Examiner's tip

It is very important to spell this word correctly. It is very similar to the other type of cell division, meiosis, and so if you spell it incorrectly you are likely to be marked wrong.

c) Identical **1**
Genetic copy **1**

Examiner's tip

Most people know that a clone is an identical copy but you need to point out that it is genetically identical in order to score full marks.

d) Quicker
All identical so can predict what the plants will look like
Copy attractive plant
Only need one parent any two **2**

4 a) All chromosomes in pairs apart from two **1**
Males have one X and one Y chromosome **1**
b) They would not form pairs **1**
Only 23 would be present **1**

Examiner's tip

Sometimes examiners might give you a diagram like this but with an extra chromosome, i.e. 47 in total. Conditions such as Down's syndrome are caused by this.

c) i) Meiosis **1**
ii) Mutations **1**

Examiner's tip

Make sure that you can give one environmental cause of mutations, such as high energy radiation.

d) i) Changing the genes of an organism **1**
By moving DNA from one organism to another **1**
ii) Ethical objections
Unforeseen disadvantages
May be misused by unscrupulous individuals **3**

CHAPTER 5

Living things in their environment

To revise this topic more thoroughly, see Chapter 5 in *Letts Revise GCSE Science Study Guide.*

Try this sample GCSE question and then compare your answers with the Grade C and Grade A model answers on the next page.

Pesticides are chemicals that are often sprayed on crops.
The diagram shows a farmer spraying his crops in a field near the sea.

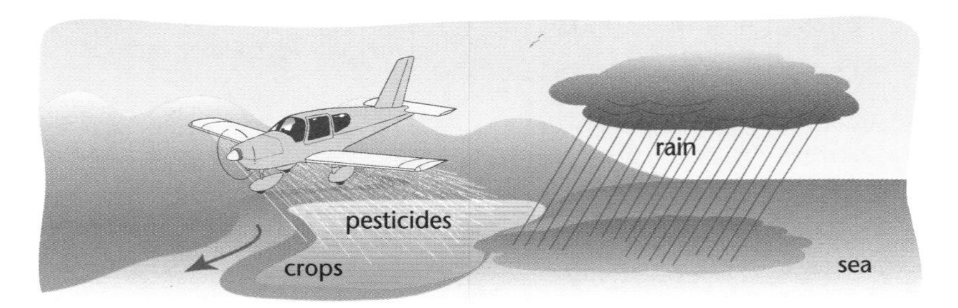

a **(i)** Why does the farmer spray his crops with pesticide? **[2]**

(ii) Some of the pesticide sprayed by the farmer can be found in the sea, even though the pilot is careful not to spray the sea directly.

Explain how the pesticide enters the sea. **[2]**

b This diagram shows a food chain found in the sea.

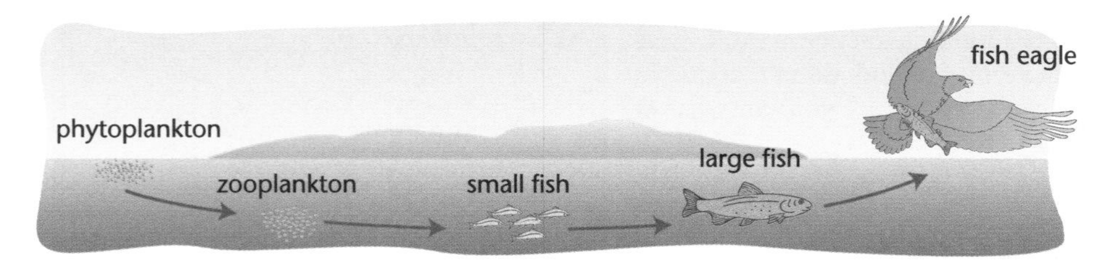

(i) Draw a pyramid of biomass for this food chain. **[2]**

(ii) After a few years the local people started to notice that the numbers of fish eagles were decreasing although the numbers of other organisms in the chain were not going down.

Explain how the pesticide may have caused this. **[4]**

(Total 10 marks)

These two answers are at Grades C and A. Compare which one your answer is closest to and think how you could have improved it.

GRADE C ANSWER

Athel

a (i) To kill pests such as insects in the field. ✓

(ii) The pesticide passes into the sea in rivers. ✓

b (i) ✓✓

(ii) Pesticides are poisonous chemicals. ✓ The eagle is poisoned by the pesticides causing eutrophication in the sea. ✗

5 marks = Grade C answer

Athel scores one mark but his answer misses the second point.

Athel scores one mark here but by not mentioning the rain, he fails to use all the information on the diagram.

This time the information on the diagram helps Athel to score both marks. Correct shape and five levels.

This is the hardest part of the question. Athel scores one mark for realising that pesticides are poisonous chemicals but he then confuses pesticides with fertilisers.

Grade booster ···> move a C to a B

Athel needs to use all the information given in the question. By commenting on the rain in a (ii) and answering fully in a (i) he could have increased his grade by one.

GRADE A ANSWER

Natasha

a (i) The farmer uses pesticide to kill insects on his crops ✓ and so increases his yield. ✓

(ii) The pesticide gets leached into rivers by rain, ✓ it then flows into the sea. ✓

b (i) ✓✓

(ii) The pesticide is taken in by the plankton but the amounts are too small to poison them. ✓ The pesticide is passed through the food chain and is taken in by the eagle in its food. ✓ The pesticides are poisonous and kill the eagle. ✓

9 marks = Grade A answer

Unlike Athel, Natasha has explained why the farmer needs to kill pests.

Natasha's answer uses the correct term leaching, although this is not necessary to score full marks.

Again full marks are scored here. Correct shape and five levels.

This is a much better answer than Athel's. Natasha explains why the smaller organisms are not killed. To get full marks she needs to explain that the pesticide builds up in concentration in the larger animals.

Grade booster ···> move A to A*

Natasha scores nine marks from a possible ten. The only mark that she has lost is for failing to explain why the pesticide kills the eagle. The principle of the pesticide building up higher up food chains is a difficult idea to understand.

QUESTION BANK

1 The graph shows changes in the carbon dioxide level of the Earth's atmosphere.
It also shows how the temperature of the atmosphere has differed compared to today's temperature.
The changes are shown from 160 thousand years ago up to the present day.

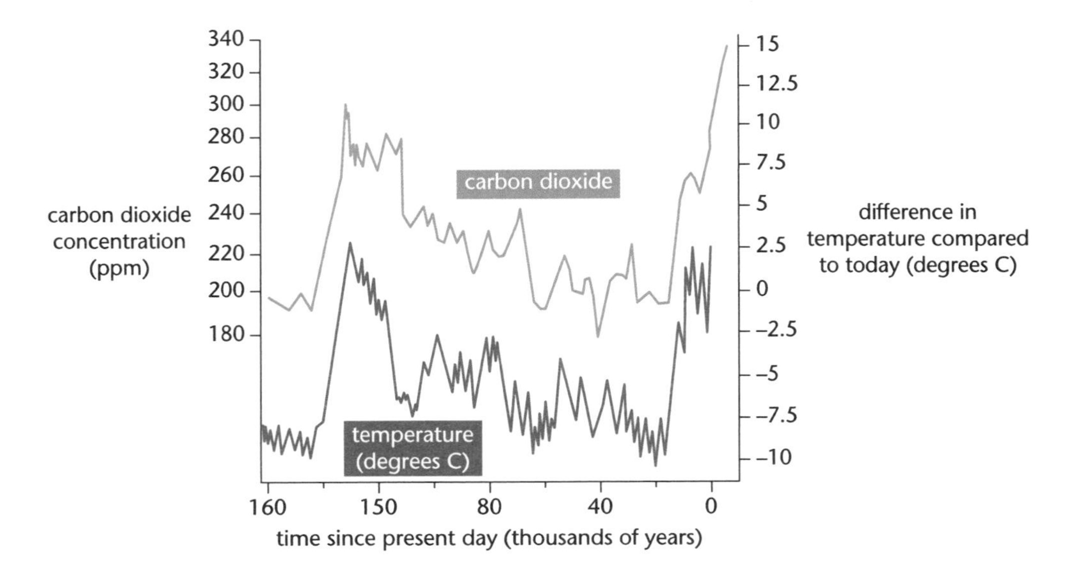

a) Describe how the temperature of the atmosphere has differed in the past compared to today's temperature.

....

....

.... (3)

b) Some scientists think that there is a link between carbon dioxide levels and the temperature of the Earth.

i) What evidence is there in this graph to support the scientists' ideas?

....

....

.... (2)

ii) Explain how changes in carbon dioxide levels may cause a change in the temperature of the Earth.

....

....

....

.... (4)

c) People are trying to stop carbon dioxide levels from rising.
Write down two different ways that they are using.

1 ....

....

2 ....

.... (2)

TOTAL 11

ANSWERS ON PAGE 41 ANSWERS ON PAGE 41 ANSWERS ON PAGE 41 ANSWERS ON PAGE 41

2 The diagram shows part of the nitrogen cycle.

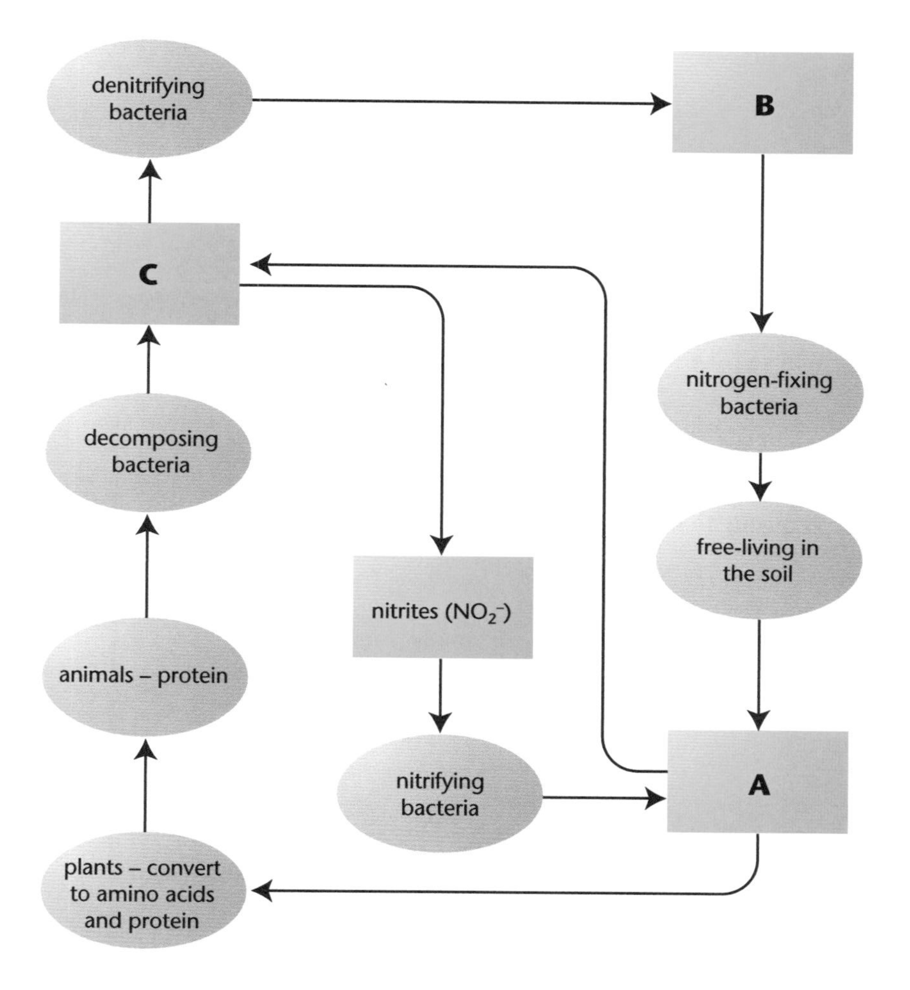

a) The letters **A, B** and **C** represent three chemicals that contain nitrogen. Write down the names of these three chemicals.

A = ...

B = ...

C = ... (3)

b) i) The cycle shows bacteria decomposing dead material. Name one other type of organism that decomposes dead material.

.... (1)

ii) For decomposers to break down dead material rapidly they need ideal environmental conditions. Write down three of these ideal conditions.

1 ....

2 ....

3 .... (3)

c) The nitrogen-fixing bacteria shown on the diagram live free in the soil. Describe another place where nitrogen-fixing bacteria can be found.

....

.... (2)

TOTAL 9

ANSWERS ON PAGE 41 ANSWERS ON PAGE 41 ANSWERS ON PAGE 41 ANSWERS ON PAGE 41

3 Algae are microscopic plants that live in ponds.
The graph shows the number of algae living in a pond at different times of the year.

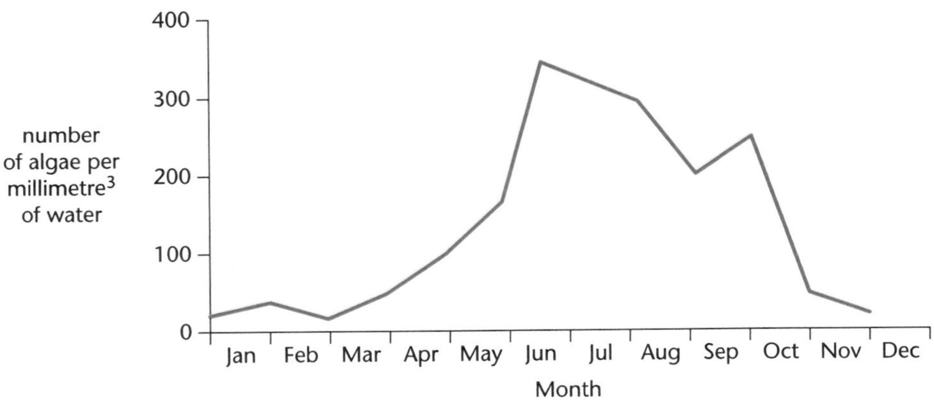

a) i) In which month are algae numbers at their highest and what is their maximum number?

Month ....

Maximum number .... per millilitre (2)

ii) The number of algae is starting to increase in March.
Suggest why this is.

....

....

.... (2)

b) Small animals in the pond feed on the algae.
Draw a line on the graph to show how the number of small animals would vary throughout the year. (2)

c) A number of water lily plants were introduced into the pond.
They have large leaves that float on the surface of the pond.
Suggest what effect the introduction of water lilies would have on the number of algae in the pond.

Explain your answer.

Effect ....

Explanation ....

....

.... (3)

d) Some fertiliser from a nearby farm was washed into the pond.
After a while it caused many of the large fish in the pond to die.
Explain how the fertiliser may have caused the death of the fish.
One mark awarded for a clearly ordered answer.

....

....

....

....

.... (4)+(1)

TOTAL 14

ANSWERS ON PAGE 41 ANSWERS ON PAGE 41 ANSWERS ON PAGE 41 ANSWERS ON PAGE 41

QUESTION BANK ANSWERS

1 a) The temperature has generally been colder than today (1)
Apart from a period about 152 thousand years ago (1)
Temperature started to warm up about 15 – 20 thousand years ago (1)

b)i) The two graphs follow the same pattern (1)
Both have peaks at 152 thousand years ago/both have increased since 15 – 20 thousand years ago (1)

ii) Carbon dioxide lets through short wave radiation from the sun (1)
Earth gives out long wave radiation (1)
This is prevented from leaving the atmosphere by carbon dioxide (1)
Reference to global warming or the greenhouse effect (1)

Examiner's tip

Many candidates could write down the phrase 'the greenhouse effect', but without any explanation this will only score one mark.

c) Use of alternative energy sources or example such as wind power
Recycling of materials or example such as bottle banks
Details of a scheme to use less energy such as low energy bulbs — any two (2)

Examiner's tip

The question asks for two different ways of reducing carbon dioxide output.
You will not score two marks by giving two similar answers such as 'turning off lights' and 'turning off the television' when they are not needed.

2 a) **A** = nitrates (1)
B = nitrogen (gas) (1)
C = ammonium compounds (1)

b)i) Fungi (1)

ii) Warm temperature
Moisture
Oxygen
Suitable pH — any three (3)

Examiner's tip

The question asks for the ideal conditions.
Do not write simply 'temperature' or 'pH'.

c) In root nodules (1)
Of leguminous plants (1)

Examiner's tip

The word 'describe' in the question and the two marks available should tell you that the examiner is looking for more than just 'root nodules' in an answer.

3 a) i) Month = June (1)
Maximum number = 350 per millilitre (1)

ii) More sunlight (1)
More photosynthesis (1)

b) Line drawn to show similar pattern to algae numbers (1)
Line drawn to peak just after June (1)

Examiner's tip

Adding this line makes this graph a predator–prey graph. The graph for the predators usually peaks just after the peak of the prey. Make sure you can explain why. Usually the numbers of the predators (the small animals) would be much lower than the prey.

c) Algae numbers would decrease (1)
More competition for light/water lilies block the light (1)
Less photosynthesis so less food produced by algae (1)

Examiner's tip

This question is all about competition. The algae and the water lilies could also be competing for minerals.

d) Fertiliser stimulates the growth of the algae
Large numbers of algae die
Bacteria decompose the algae
Bacteria use up the oxygen
Fish cannot respire and so die
Reference to eutrophication — any four (4)
plus one mark for a clearly ordered answer (1)

Examiner's tip

The idea of eutrophication is often explained incorrectly in answers. Many candidates say that it is the algae that use up all the oxygen rather than the decomposing bacteria. Other candidates get confused between fertilisers and pesticides.
The extra mark for quality of written communication is given if you use correct scientific terms and link them together in your answer.

CHAPTER 6

Classifying materials

To revise this topic more thoroughly, see Chapter 6 in *Letts Revise GCSE Science Study Guide*.

Try this sample GCSE question and then compare your answers with the Grade C and Grade A model answers on the next page.

a All atoms, except for a simple hydrogen atom, are made up of three different particles.

(i) Write down the names, relative charges and relative masses of these particles. **[3]**

(ii) Describe where these particles are found in a typical atom. **[2]**

b Oxygen exists as a mixture of three isotopes.

$$^{16}_{8}O \quad ^{17}_{8}O \quad ^{18}_{8}O$$

These three isotopes contain different numbers of particles.

(i) Which particles are present in the same number in each isotope? **[1]**

(ii) Which particle is present in different numbers in each isotope? **[1]**

c About 100 years ago scientists thought that atoms had a uniform density.

Ernest Rutherford carried out an experiment.

He found that when he fired high energy particles at thin gold foil, most of the particles went straight through without being deflected.

About 1 in every 10 000 was greatly deflected, sometimes through nearly 180°.

Why did these experiments of Rutherford change the thoughts about atomic structure? **[3]**

(Total 10 marks)

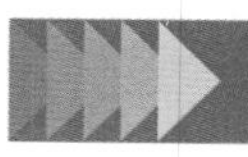
These two answers are at Grades C and A. Compare which one your answer is closest to and think how you could have improved it.

GRADE C ANSWER

Simon makes two comments that are correct but are not complete. The statement about charge does not include neutrons and does not make it clear that the charge on the proton and the electron is 1. The statement about mass does not make it clear that the proton and neutron have the same mass.

Simon

a (i) Protons, neutrons and electrons ✓
Protons are positive and electrons are negative. Protons and neutrons are heavier than electrons. ✗✗

(ii) Protons and neutrons are in the nucleus ✓
Electrons around the nucleus ✓

b (i) Protons and electrons ✓

(ii) Neutrons ✓

c

✗✗✗

This is very difficult and Simon has made no attempt to answer it.

5 marks = Grade C answer

Grade booster ····› move a C to a B
If Simon had been more exact in his answers to a (i) it could have boosted his grade. Also perhaps he could have had an attempt at c.

GRADE A ANSWER

This is an excellent answer. Putting the information in a table was a good idea. It makes sure that nothing is missed.

Sabil

a (i)

particle	charge	mass
proton	+1	1 unit
electron	-1	negligible
neutron	0	1 unit

✓✓✓

(ii) Protons and neutrons are in the nucleus ✓
Electrons around the nucleus ✓

b (i) Protons and electrons ✓ (ii) Neutrons ✓

c If gold atom had uniform density all particles would pass through without deflection. ✓ Particles are deflected when they hit a heavy part of the atom. ✓

A question like this is testing understanding of atomic structure. Sabil fails to link the observations with the existence of a nucleus at the centre of the atom where all the mass is concentrated.

9 marks = Grade A answer

Grade booster ····› move A to A*
Sabil has only made two points in a question with three marks allocated. He should have looked for another point. Any mention of nucleus would have been sufficient.

Try this sample GCSE question and then compare your answers with the Grade C and Grade A model answers on the next page.

The diagrams show the electron arrangements in hydrogen, sodium and fluorine atoms.

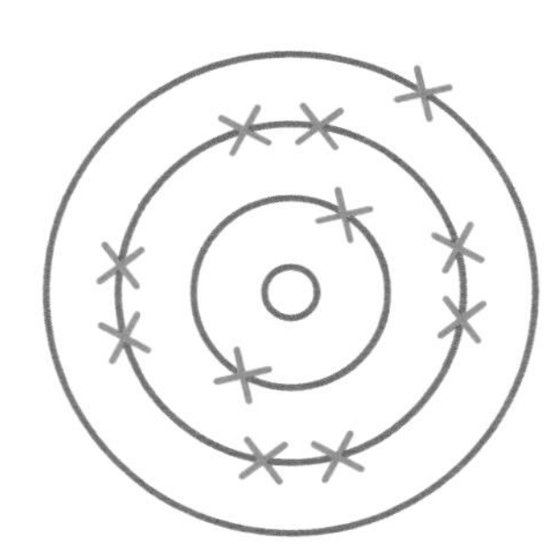

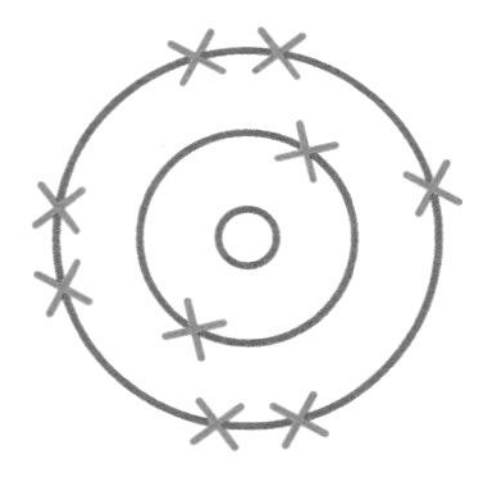

a Sodium and fluorine combine to form solid sodium fluoride, NaF.

(i) What type of bonding is present in sodium fluoride? **[1]**

(ii) Describe the changes that take place when sodium and fluorine atoms combine. **[3]**

b Hydrogen fluoride, HF, is a gas.

(i) What type of bonding is present in hydrogen fluoride gas? **[1]**

(ii) Draw a dot and cross diagram for a molecule of hydrogen fluoride. **[2]**

(iii) When hydrogen fluoride dissolves in water, the solution produced conducts electricity. What change of bonding explains this? Explain your answer. **[2]**

(Total 9 marks)

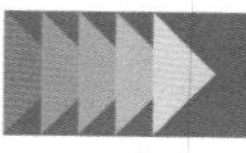

These two answers are at Grade C and A. Compare which one your answer is closest to and think how you could have improved it.

GRADE C ANSWER

*Choosing covalent bonding was a costly mistake. Zara should have realised that the bonding involved a metal atom and a non metal atom. This is usually ionic. The wrong answer to **(ii)** is a consequence of the answer to **(i)**. It is totally wrong and the examiner cannot allow any credit.*

Zara

a (i) Covalent ✗ **(ii)** Sodium and fluorine atoms each provide an electron and the pair of electrons is shared. ✗✗✗

b (i) Covalent ✓

(ii)

✓✓

(iii) Bonding changes from covalent to ionic ✓✗

Zara should have linked the solution conducting electricity with the presence of hydrogen and fluoride ions.

4 marks = Grade C answer

Grade booster ····> move a C to a B

Zara's answer is just Grade C. She falls down with the harder parts a and b. Many candidates do. She should have used ideas of electron arrangement. Look at the information given and the answers you have already made. They can often help with harder parts of the question.

GRADE A ANSWER

This is very good answer showing a good understanding of bonding.

Georgina

a (i) Ionic ✓

(ii) A sodium atom loses an electron ✓ and a fluorine atom gains an electron. ✓ The ions are held together by electrostatic forces. ✓

b (i) Covalent ✓

(ii)

Both students have chosen to show only the outer shell of electrons. This is quite acceptable as bonding only involves outer electrons. Showing all shells can sometimes complicate the diagram.

Georgina fails to distinguish electrons from the hydrogen atom and electrons from the fluorine atom. If you look back, Zara used a dot for the electron from the hydrogen and a cross for electrons from the fluorine atom. It is clear from Zara's diagram that one electron comes from the hydrogen atom and one from the fluorine atom. It is not possible to do this from Georgina's.

✓✗

(iii) Bonding changes from covalent to ionic. ✓ The solution contains H^+ and F^- ions which conduct electricity. ✓

8 marks = Grade A answer

Grade booster ····> move A to A*

Georgina's answer was very close to being perfect. It involved just a slight slip in using crosses for all electrons. She should just have checked this before moving on. It is a slip that is easy to make.

QUESTION BANK

1 Substances can have different structures.
These are:

covalent giant structure **covalent molecular structure**

metallic **ionic giant structure**

The table gives some information about four substances.
Complete the table.

Substance	Appearance	Melting point	Electrical conductivity		Structure
			When solid	When liquid	
copper	red-brown solid	high	good	good	metallic
potassium chloride	white crystals	high			
iodine	dark grey solid	low			
silicon dioxide	white solid	high			

(6)

TOTAL 6

---

2 Boron is a non-metallic element.

a) The table shows information about two types of boron atom.

Atom	Symbol	Protons	Electrons	Neutrons
boron-10	$^{10}_{5}B$			
boron-11	$^{11}_{5}B$			

i) Finish the table. (3)

ii) What scientific word is used for different atoms of the same element with different mass numbers?

.... (1)

iii) The relative atomic mass of boron measured precisely is 10.82.
What does this tell you about the relative amounts of the two different boron atoms in a sample of boron? (No calculation required.)

....

.... (2)

iv) What is the electron arrangement in a boron atom?

.... (1)

ANSWERS ON PAGE 49 ANSWERS ON PAGE 49 ANSWERS ON PAGE 49 ANSWERS ON PAGE 49

b) The 'dot and cross' diagram shows the structure of a molecule of boron chloride.

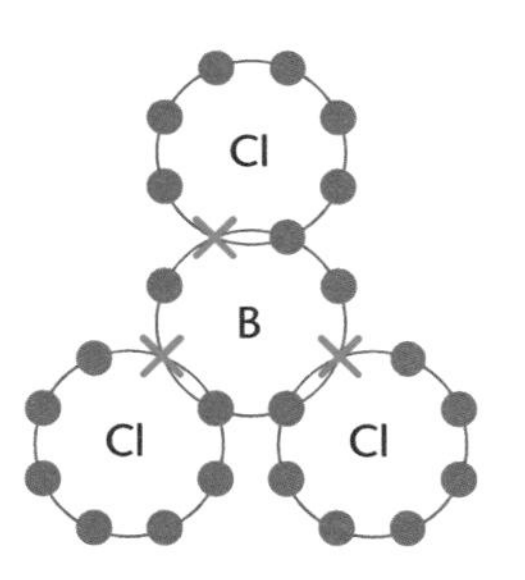

Describe the changes in electron arrangement which occur when boron chloride is formed from boron and chlorine atoms.

....

.... (2)

c) The table shows the melting points of some boron compounds.

Compound	Melting point in °C	Boiling point in °C
boron oxide, B_2O_3	460	1860
boron hydride, B_2H_6	−165	−92
boron chloride, BCl_3	−107	12
boron bromide, BBr_3	−50	91

i) What does the information tell you about the structures of boron oxide and boron hydride?

....

.... (2)

ii) Explain why the boiling point of boron bromide is higher than the boiling point of boron chloride.

....

.... (2)

TOTAL 13

3 Beryllium has a mass number of 9 and an atomic number of 4.

a) How many protons are there in a beryllium atom?

.... (1)

b) How many electrons are there in a beryllium atom?

.... (1)

c) How many neutrons are there in a beryllium atom?

....

.... (1)

d) Draw a diagram to show the arrangement of particles in a beryllium atom. Show on the diagram where protons, neutrons and electrons are. (4)

TOTAL 7

ANSWERS ON PAGE 49 ANSWERS ON PAGE 49 ANSWERS ON PAGE 49 ANSWERS ON PAGE 49

4 Diamond and graphite are two allotropes of carbon.

a) Describe the structure of diamond and graphite.

....

....

....

.... (4)

b) Write down one property of each allotrope and explain how the structure of the allotrope is consistent with this property.

Property of diamond ....

Explanation ....

....

Property of graphite ....

Explanation ....

.... (4)

c) CFCs are compounds of carbon, chlorine and fluorine. One such compound is dichlorodifluoromethane, CF_2Cl_2. This has a melting point of −158°C and a boiling point of −30°C.

i) What type of structure does dichlorodifluoromethane have?

.... (1)

ii) What type of bonding is present in dichlorodifluoromethane?

.... (1)

TOTAL 10

---

5 The electron arrangements of calcium and oxygen atoms are:
Calcium 2,8,8,2 Oxygen 2,6

a) i) Describe the electron changes that occur when calcium and oxygen combine.

....

.... (2)

ii) Explain why calcium oxide has a very high melting point.

....

.... (2)

iii) What type of bonding is present in calcium oxide?

.... (1)

b) i) Draw a 'dot and cross' diagram for an oxygen, O_2, molecule.
You need only show electrons in the outer shell. (2)

ii) What type of bonding is present in oxygen? .... (1)

TOTAL 8

ANSWERS ON PAGE 49 ANSWERS ON PAGE 49 ANSWERS ON PAGE 49 ANSWERS ON PAGE 49

QUESTION BANK ANSWERS

1

Substance	Appearance	Melting point	Electrical conductivity		Structure
			When solid	When liquid	
copper	red-brown solid	high	good	good	metallic
potassium chloride	white crystals	high	poor	good	ionic giant structure
iodine	dark grey solid	low	poor	poor	covalent molecular structure
silicon dioxide	white solid	high	poor	poor	covalent giant structure

There are 6 gaps for electrical conductivity.
You score 1 mark for each two correct answers.
There is one mark for each structure.
Maximum marks 6

2 a) i) Boron-10 5 protons 5 electrons 5 neutrons
Boron-11 5 protons 5 electrons 6 neutrons 3
all correct – 3 marks, 4 or 5 correct – 2 marks, 3 or 2 correct – 1 mark

ii) Isotope 1

iii) The relative atomic mass of boron is closer to 11 than 10 1
There must be much more boron-11 than boron-10. 1

iv) 2,3 1

Examiner's tip

If you gave the number of electrons in a boron atom as 6, for example, the electron arrangement 2,4 would be consistent and you would be given this mark for error carried forward. However, from the position in the Periodic Table (group 3), you should realise that there are three electrons in the outer shell.

b) Electrons from boron and chlorine atoms 1
Are shared 1

Examiner's tip

You must realise that the bonding in boron chloride is covalent and not ionic. Ionic bonding involves a metal and a non-metal. Here we have two non-metals, boron and chlorine. Also molecules are formed and not ions.

c) i) Boron oxide – giant structure 1
Boron hydride – molecular structure 1

Examiner's tip

The information in the table can be used to decide on the structure of these compounds. Boron oxide is a solid and has a relatively high melting point. This does not tell us whether it is a giant structure of atoms or a giant structure of ions. We could decide this by testing to see whether the melt conducts electricity or not. If the melt conducts electricity, the original solid has a giant structure of ions. If it does not, it has a giant structure of atoms.
Boron hydride is a gas and has a low melting point and low boiling point.

ii) Bromine is heavier than chlorine, so boron bromide molecules are heavier than boron chloride molecules. 1
So more energy is needed to make it boil/become a gas/separate molecules. 1

3 a) 4 1
b) 4 1
c) 5 1

Examiner's tip

The number of protons is the same as the atomic number. The number of electrons is the same as the number of protons. The number of neutrons is the difference between the mass number and the atomic number.

FOR MORE INFORMATION ON THIS TOPIC ... SEE REVISE GCSE SCIENCE ... CHAPTER 6

d) 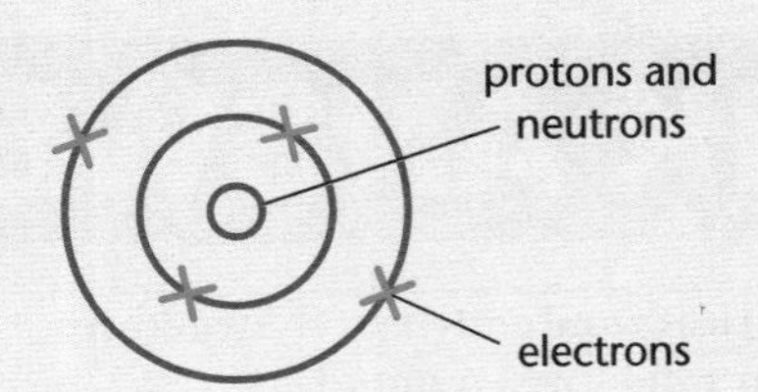

1 mark is for indicating that protons and neutrons are in the nucleus in the centre of the atom; 1 mark is for showing electrons in shells around nucleus; 1 mark for two electron shells;
1 mark for electron arrangement 2,2 ❹

❹ a) Diamond – all atoms joined by strong (covalent) bonds ❶
produces three dimensional structure ❶
Graphite – strong (covalent) bonds within layer ❶
very weak forces between layers ❶
a diagram of each structure alone is worth two marks

b) Very hard ❶
All atoms bonded together so bonds have to be broken. ❶
Good lubricant ❶
Layers are able to slide over each other ❶

c) i) Molecular ❶
ii) Covalent ❶

❺ a) i) Each calcium atom loses two electrons ❶
Each oxygen atom gains two electrons ❶

Examiner's tip

It is important to stress that in ionic bonding electrons are completely transferred and not just shared.

ii) Ions produced have 2+ and 2– charges ❶
A large amount of energy needed to break structure ❶

iii) Ionic ❶

b) i) 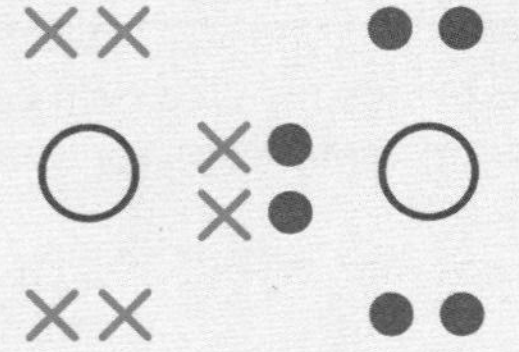

one mark for two pairs of electrons between two oxygen atoms ❷

ii) Covalent ❶

Classifying materials

FOR MORE INFORMATION ON THIS TOPIC ... SEE REVISE GCSE SCIENCE ... CHAPTER 6

CHAPTER 7

Changing materials

To revise this topic more thoroughly, see Chapter 7 in *Letts Revise GCSE Science Study Guide.*

Try this sample GCSE question and then compare your answers with the Grade C and Grade A model answers on the next page.

Decane is an alkane containing 10 carbon atoms. It is a liquid.
When decane vapour is passed over heated broken china, decane is broken down by a process of cracking.

a When decane vapour is cracked a mixture of products is formed, shown in the equation:

$$C_{10}H_{22} \rightarrow C_2H_4 + C_2H_6$$

(i) Balance the equation. **[1]**

(ii) Write down the names of the two products. **[2]**

(iii) Describe a test that could be used to distinguish the two products. **[2]**

b C_2H_6 is used as a fuel.

(i) Write a balanced symbol equation for the combustion of this gas in excess air. **[3]**

(ii) Why is it dangerous if there is a limited amount of air available? **[2]**

c The product C_2H_4 is used to make a polymer. It can also be used to make a range of carbon compounds.

(i) Draw the structure of the C_2H_4 molecule. **[1]**

(ii) Draw the structure of the molecule formed when water is added to C_2H_4. **[1]**

(Total 12 marks)

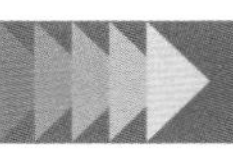

These two answers are at Grades C and A. Compare which one your answer is closest to and think how you could have improved it.

GRADE C ANSWER

David

a (i) $C_{10}H_{22} \rightarrow 5C_2H_4 + H_2$ ✗

David has made the mistake of changing the products in the equation. The equation written is balanced but is not the one required.

(ii) Ethene ✓ ethane ✓

(iii) Ethane is saturated, ethene is unsaturated ✗ ✗

The statement given is correct but it is not a test. David is not answering the question.

b (i) C_2H_6 + air $\rightarrow CO_2 + H_2O$ ✗✓✗

David does not realise that oxygen is the reactant. The first mark for the reactants is not given but the mark for the products can still be awarded. It is not possible to score the third mark for balancing.

(ii) If there is a limited amount of air, carbon monoxide may be formed. ✓ This is poisonous. ✓

This is a good answer.

c (i)

```
H       H
 \     /
  C = C
 /     \
H       H
```

✓

(ii)

```
H       H
 \     /
  C - C - H - O - H
 /     \
H       H
```

✗

*The structure of ethene is correct but the product in **c (ii)** is incorrect. One carbon atom only forms three bonds instead of four.*

6 marks = Grade C answer

Grade booster ····> move a C to a B

Make sure the answer given is to the question asked and not to a similar one. A test requires a description of the test and the result.

GRADE A ANSWER

Callum

a (i) $C_{10}H_{22} \rightarrow 4C_2H_4 + C_2H_6$ ✓

(ii) ethene ✓ ethane ✓

(iii) Add bromine solution
Ethene decolourises bromine ✓

*In **a (iii)** Callum implies that as ethene decolourises bromine, ethane does not. To be sure of the mark here he should state what happens to both.*

b (i) $C_2H_6 + O_2 \rightarrow CO_2 + 3H_2O$ ✓✓✗

*In **b (i)** the equation is not correctly balanced and so the third mark is lost.*

(ii) If there is a limited amount of air, carbon monoxide may be formed. ✓
This is poisonous. ✓

c (i)

```
H       H
 \     /
  C = C
 /     \
H       H
```

✓

(ii)

```
    H   H
    |   |
H - C - C - O - H
    |   |
    H   H
```

✓

Full marks are scored in the rest of the question.

10 marks = Grade A answer

Grade booster ····> move A to A*

This question confirms the importance of equation writing and being able to draw correct structures. The correct equation is $2C_2H_6 + 7O_2 \rightarrow 4CO_2 + 6H_2O$

Try this sample GCSE question and then compare your answers with the Grade C and Grade A model answers on the next page.

Iron is extracted from iron ore in a blast furnace.

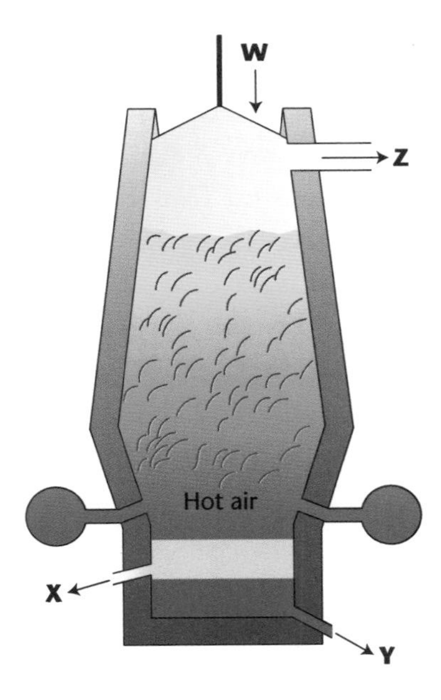

a Write down the materials added to the furnace at W. **[2]**

b What is tapped off at X and at Y? **[2]**

c In the furnace, carbon monoxide is the active reducing agent.
Describe how carbon monoxide is formed in the furnace. **[2]**

d The gases given off from the furnace are called waste gases.
Suggest two gases, apart from carbon monoxide, that will be present in large amounts in these gases. **[2]**

e The equation for the reduction of iron(III) oxide is

$$Fe_2O_3 + 3CO \rightarrow 2Fe + 3CO_2$$

Calculate the maximum mass of iron that could be produced from 16 tonnes of iron(III) oxide.
(Relative atomic masses: Fe = 56; O = 16) **[3]**

(Total 11 marks)

These two answers are at Grade C and A. Compare which one your answer is closest to and think how you could improve it.

GRADE C ANSWER

Nazim

a Iron ore, coal and limestone ✗ ✓

b X iron Y slag ✗ ✓

c Coal burns to form carbon dioxide ✓ Carbon dioxide reacts with carbon to form carbon monoxide ✓

d Carbon dioxide ✓ oxygen ✗

e RFM of Fe_2O_3 = (2 x 56) + (3 x 16) = 160 ✓

Coal is not an alternative to coke. There is one mark for two correct raw materials.

The two products are correct but the wrong way round. This scores one mark out of two. Slag is less dense than molten iron and so floats on it.

Oxygen is not in the waste gases.

Nazim scores only the first mark in the calculation.

6 marks = Grade C answer

Grade booster ····> move a C to a B

Chemical calculations are important on higher tier papers. Practice with calculations of this type would improve Nazim's grade.

GRADE A ANSWER

Dinah

a Haematite, coke and limestone ✓ ✓

b X slag Y iron ✓ ✓

c $C + O_2 \rightarrow CO_2$ ✓
$C + CO_2 \rightarrow 2CO$ ✓

d Nitrogen ✓ argon ✗

e RFM of Fe_2O_3 = (2 x 56) + (3 x 16) = 160 ✓
160 tonnes produce 112 tonnes ✓
16 tonnes produce 11.2 tonnes ✓

Dinah has given the correct name for an iron ore. This is more than the examiner expected.

The two correct equations are a good way of scoring these two marks.

Argon will be in the waste gases but not in large amounts.

Full marks given for the calculation.

10 marks = Grade A answer

Grade booster ····> move A to A*

Dinah's answer was very clear and concise. Her only error was in the composition of waste gases from the furnace.

QUESTION BANK

1 Crude oil is a mixture of hydrocarbons.
The diagram shows how this raw material is split up into useful products in a petrochemical plant.

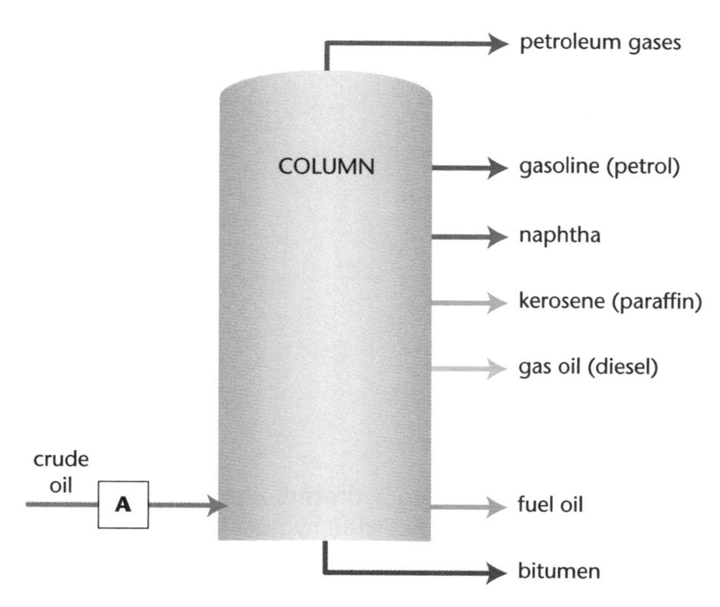

a) Write down the name of the process shown in the diagram.

.... ①

b) What is happening at **A** in the diagram?

.... ①

c) State two differences in properties between gasoline and fuel oil.

1 ....

2 .... ②

d) The flow diagram shows what happens to most of the naphtha.

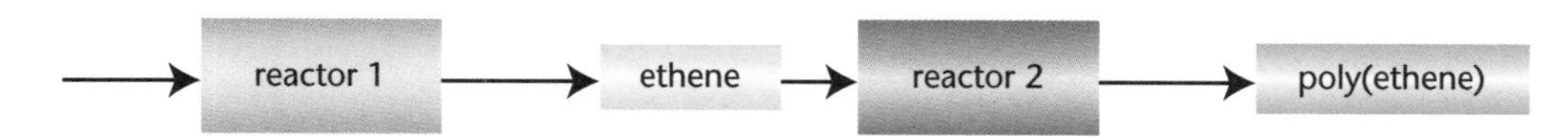

i) What type of reaction is taking place in reactors 1 and 2?

Reactor 1 ....

Reactor 2 .... ②

ii) Suggest why it is economic to use naphtha in this way.

....

.... ①

ANSWERS ON PAGE 61 ANSWERS ON PAGE 61 ANSWERS ON PAGE 61 ANSWERS ON PAGE 61

e) The petroleum gases contain methane, ethane, C_2H_6 and propane, C_3H_8.
The structure of a molecule of ethane is shown below

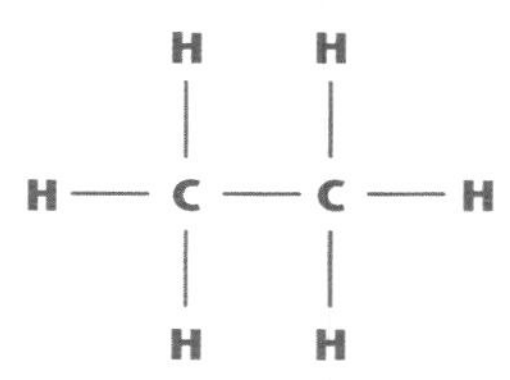

i) What is the molecular formula of methane?

.... ①

ii) Draw the structure of a molecule of propane. ①

f) Methane, ethane and propane are saturated hydrocarbons.
What does saturated mean when used to describe hydrocarbons?

.... ①

TOTAL 10

---

2 The Haber process is used in industry to produce ammonia.
The important reaction between nitrogen and hydrogen is a reversible reaction.

$$N_2(g) + 3H_2(g) \rightleftharpoons 2NH_3(g)$$

a) The graph shows the percentage of ammonia produced at different temperatures and pressures.

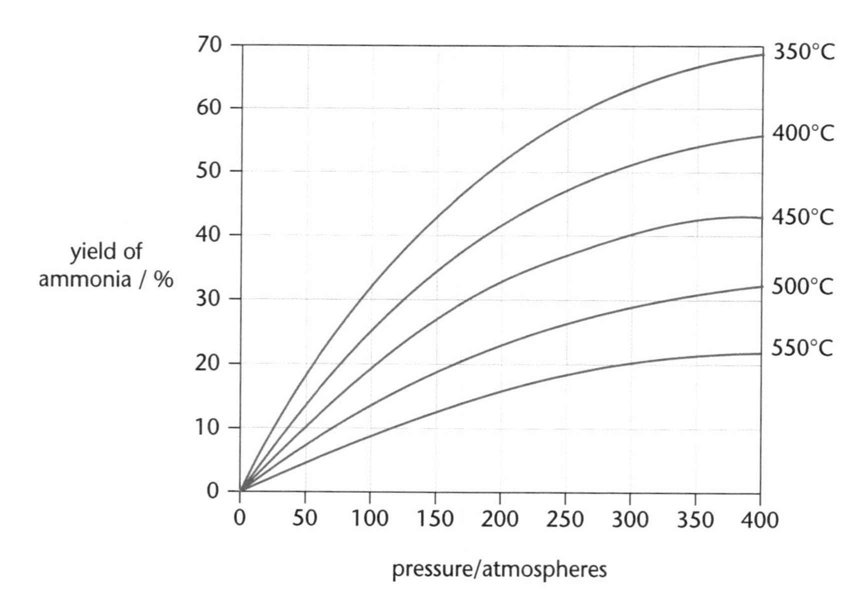

ANSWERS ON PAGE 61 ANSWERS ON PAGE 61 ANSWERS ON PAGE 61 ANSWERS ON PAGE 61

What conditions of temperature and pressure produce the highest yield of ammonia?

.... (2)

b) An iron catalyst is used in the Haber process.

i) What effect does an iron catalyst have on the rates of forward and reverse reactions?

.... (2)

ii) What effect does an iron catalyst have on the yield of ammonia at equilibrium?

.... (1)

iii) State three conditions necessary for a reversible reaction to establish 'dynamic equilibrium'.

1 ....

2 ....

3 .... (3)

c) After passing over the iron catalyst, the mixture of gases contains ammonia, hydrogen and nitrogen. The table shows the boiling point of each gas.

Gas	Boiling point in °C
ammonia	−33
hydrogen	−253
nitrogen	−196

i) Use this information to explain how the ammonia is extracted from the mixture.

....

....

.... (3)

ii) Describe what happens to the hydrogen and nitrogen after the ammonia has been extracted from the mixture.

....

.... (2)

d) Much of the ammonia produced in the Haber process is used to make fertilisers such as ammonium sulphate.

$$2NH_3(g) + H_2SO_4(aq) \rightarrow (NH_4)_2SO_4(aq)$$

i) Calculate the relative formula masses of ammonia and ammonium sulphate.
(Relative atomic masses: H = 1, N = 14, O = 16, S = 32)
Show your working.

Ammonia ....

Ammonium sulphate .... (2)

ii) Calculate the mass of ammonium sulphate produced from 17 tonnes of ammonia.

Answer = ... tonnes (2)

TOTAL 17

ANSWERS ON PAGE 61 ANSWERS ON PAGE 61 ANSWERS ON PAGE 61 ANSWERS ON PAGE 61

3 During a chemistry lesson, groups of students were burning strips of metal in air. The mass of the metal and the mass of metal oxide produced were recorded. The results are shown in the table.

Group number	Mass of metal in g	Mass of metal oxide in g	Mass of oxygen combined in g
1	0.11	0.18	0.07
2	0.15	0.25	0.10
3	0.20	0.34	
4	0.25	0.41	
5	0.30	0.50	

a) Finish the final column of the table. (1)

b) Plot a graph of the mass of metal (x-axis) and the mass of oxygen combined (y-axis). Draw the best straight line through the points and the origin. (3)

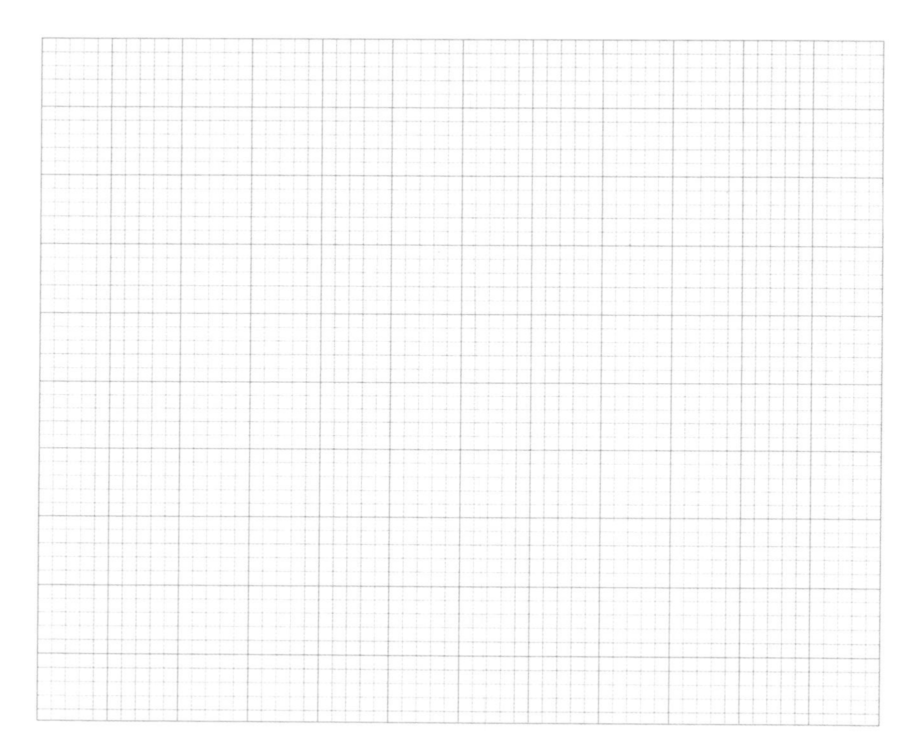

c) From the graph find the mass of oxygen which would combine with 0.24 g of metal.

Mass = ... g (1)

d) The relative atomic mass of the metal is 24 and oxygen 16.
Work out the simplest formula for the oxide.
(Use M as the symbol for the metal.) (3)

TOTAL 8

ANSWERS ON PAGE 61 ANSWERS ON PAGE 61 ANSWERS ON PAGE 61 ANSWERS ON PAGE 61

4

a) Write down the names of the three gases present in dry air in the largest amounts.

.... ③

b) Read the passage about air pollution in cities and answer the questions which follow.

Many cities around the world are suffering due to pollution from motor vehicles. Cairo (Egypt) and Santiago (Chile) are two cities in the world where air pollution is a particular problem.

In these two cities there are a large number of old and poorly-maintained cars.

Also the average speed of cars is low due to heavy traffic. The cities are in deep valleys so that pollution is not blown away. High temperatures and, in the case of Cairo, low rainfall make the problem worse.

Using old cars means the fuel must be leaded rather than the unleaded used by modern cars. Unleaded fuel produces less harmful emissions.

The chart shows the amount of the three major pollutants produced in Cairo by private cars, taxis, trucks and motorcycles.

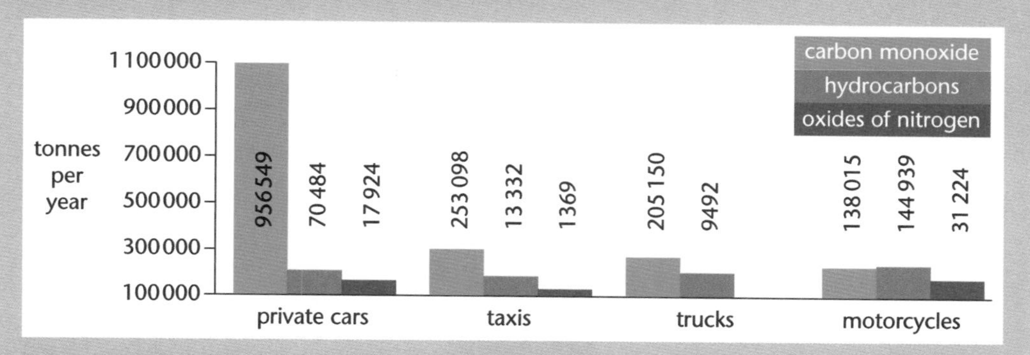

Ozone is a colourless gas produced by complex atmospheric reactions between hydrocarbons and nitrogen oxides in sunlight.

Ozone causes significant eye irritation, aggravates breathing difficulties and reduces plant growth.

Great efforts have been made to reduce air pollution. One way is to reduce the amount of traffic. In Santiago, when pollution is high, some vehicles are banned from use. For example, on a particular day vehicles with registration plates ending in 2, 4 and 6 could be banned and on another day other vehicles could be banned.

Other possible measures include introducing electric trams or battery-powered buses which do not produce pollution. Using compressed natural gas as a fuel instead of petrol or diesel reduces pollution. It produces lower carbon dioxide emissions, no sulphur dioxide, lower fuel consumption, less corrosion and less engine noise.

i) What are the three major pollutants from car engines?

.... ②

ii) Which type of vehicle in the chart produces most carbon monoxide?

.... ①

iii) Which pollutant in the passage is a secondary pollutant?

.... ①

iv) Why is ozone pollution less of a problem in cooler cities in the world?

.... ①

ANSWERS ON PAGE 61 ANSWERS ON PAGE 61 ANSWERS ON PAGE 61 ANSWERS ON PAGE 61

v) How could building ring roads around cities help air pollution?

.... ①

vi) How could planting large numbers of trees improve air pollution?

....

.... ②

TOTAL 11

---

5 Describe how aluminium is extracted from aluminium oxide in industry.
Plus 1 mark awarded for a clearly ordered answer.

....

....

....

....

....

....

....

.... ⑥+①

TOTAL 7

---

6 The diagram shows an intrusion of igneous rock into some layers of sedimentary rock.

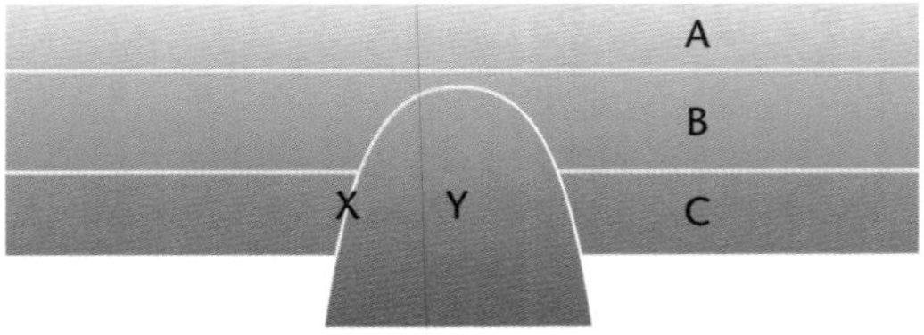

A
B
X Y C

a) How do you know that the intrusion is older than the sedimentary rock A but younger than B and C?

....

.... ②

b) The two diagrams X and Y show enlarged pictures of crystals in the igneous rock at points X and Y in the diagram above.

X Y

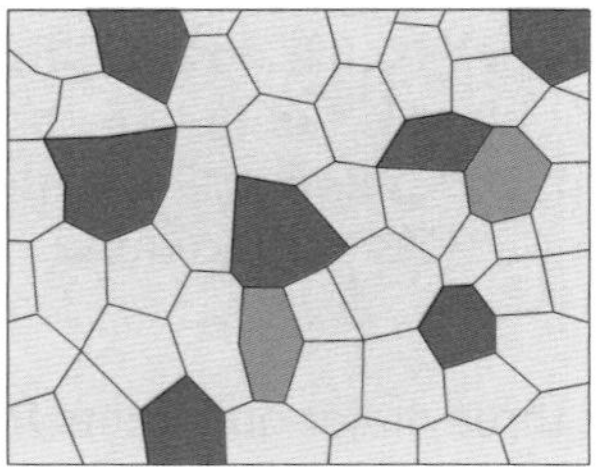

Explain why the crystals are so different in size at X and Y.

....

.... ②

TOTAL 4

ANSWERS ON PAGE 61 ANSWERS ON PAGE 61 ANSWERS ON PAGE 61 ANSWERS ON PAGE 61

QUESTION BANK ANSWERS

1 a) Fractional distillation **1**

Examiner's tip

The mark would be awarded for just distillation or fractionation.

b) Crude oil is vapourised/turned to a gas **1**

Examiner's tip

It is not enough to write that the crude oil is heated. It must be vapourised.

c) Gasoline has a lower boiling point
Is less coloured
Is easier to pour
Easier to catch alight any 2 **2**

Examiner's tip

An answer such as gasoline has shorter molecules is not acceptable as it is not a property.

d)i) Cracking **1**
Polymerisation **1**
ii) Naphtha has few uses but ethene has many uses **1**
e)i) CH_4 **1**
ii)

```
    H   H   H
    |   |   |
H — C — C — C — H
    |   |   |
    H   H   H
```
1

f) All the carbon–carbon bonds are single (covalent) bonds **1**

2 a) High pressures and low temperatures produce the highest yield of ammonia. **2**

Examiner's tip

Two marks are awarded if you give 400 atmospheres and 350°C.

b)i) A catalyst speeds up the forward reaction **1**
And speeds up the reverse reaction. **1**
ii) A catalyst does not alter the yield of ammonia **1**
iii) Closed container **1**
None of the products removed **1**
Constant temperature **1**
c)i) The mixture is cooled **1**
Below –33°C **1**
Ammonia (with a higher boiling point) liquefies. **1**
ii) The mixture of nitrogen and hydrogen is returned and mixed with the gases entering the system **1**
The gases are re-cycled **1**

d)i) Ammonia = 17 **1**
Ammonium sulphate = 132 **1**
ii) From the equation
34 tonnes of NH_3 would produce 132 tonnes of ammonium sulphate **1**
17 tonnes of NH_3 would produce 66 tonnes of ammonium sulphate **1**

3 a) 0.14 g, 0.16 g, 0.20 g all three correct **1**
b)

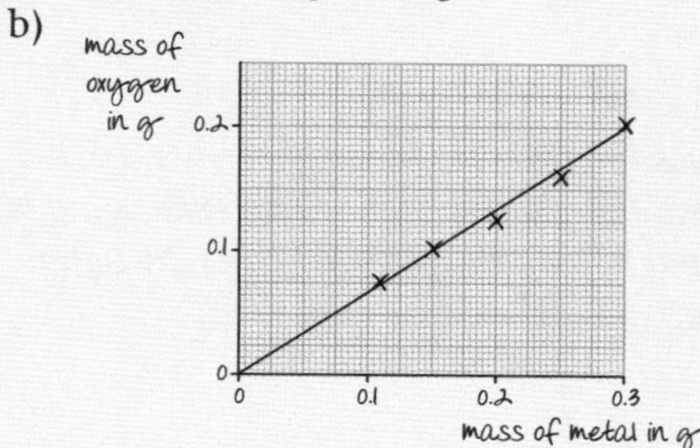

Labelling axes and using appropriate scales on graph **1**
Correct plotting **1**
Good straight line **1**

Examiner's tip

This is a typical graph on a Higher tier paper. You have to choose scales that nearly fill the grid you are given. There is only one mark for plotting and you must plot everything correctly for this. There are two anomalous points (mass of metal 0.20 and 0.25 g). Your straight line does not go through these points.
These probably occurred because not all the metal was burned or some of the oxide was lost.

c) 0.16 g **1**
d) Divide mass of metal and mass of oxygen by appropriate relative atomic masses

$\frac{0.24}{24}$ $\frac{0.16}{16}$ **1**

0.01 0.01 **1**
MO **1**

Examiner's tip

You probably realise that the metal is magnesium from the relative atomic mass given but you must not just write MO. You must show how you worked this out.

4 a) Nitrogen, oxygen and carbon dioxide 3

b) i) Carbon monoxide, hydrocarbons, oxides of nitrogen 2 marks three correct; 1 mark two correct 2

ii) Private cars 1

iii) Ozone 1

iv) Ozone is produced when sunlight acts on hydrocarbons and nitrogen oxides (there is less sunlight in cooler cities) 1

v) Air pollution is reduced when cars are moving faster. 1

vi) Photosynthesis 1

Converts carbon dioxide into oxygen 1

Examiner's tip

This question involves Ideas and Evidence. It also introduces a comprehension exercise. It is important to read the passage carefully before attempting to answer the questions.

5 Electrolysis of molten aluminium oxide.
Dissolved in molten cryolite
Carbon anodes
Aluminium produced at cathode
Oxygen produced at anode
Anodes burn away and have to be replaced
$Al^{3+} + 3e^- \rightarrow Al$
$2O^{2-} \rightarrow O_2 + 4e^-$ any 6 6
one mark for a clearly ordered answer 1

Examiner's tip

This is a good opportunity to practise extended writing. Remember 25% of your examination is for questions requiring two or more sentences. Candidates often do less well on questions requiring longer answers.

6 a) The intrusion does not cut into A so A must be younger 1

It does cut into B and C 1

b) The rate of cooling at X is much faster than at Y 1

Large crystals are formed during slow crystallisation/smaller crystals are formed during fast crystallisation 1

FOR MORE INFORMATION ON THIS TOPIC ... SEE REVISE GCSE SCIENCE ... CHAPTER 7

CHAPTER 8

Patterns of behaviour

To revise this topic more thoroughly, see Chapter 8 in *Letts Revise GCSE Science Study Guide.*

Try this sample GCSE question and then compare your answers with the Grade C and Grade A model answers on the next page.

This question is about the reaction of zinc with dilute hydrochloric acid.
One of the products is hydrogen.

a Write a balanced symbol equation for this reaction. **[3]**

b The table gives the conditions used for five experiments using zinc and dilute hydrochloric acid.

Experiment	Sample of zinc	Volume of acid in cm^3	Volume of water in cm^3	Temperature in °C	Time in min
1	1.0 g of lumps	50	0	20	3.0
2	1.0 g of powder	50	0	20	
3	1.0 g of lumps	50	50	20	
4	1.0 g of powder	50	0	40	
5	1.0 g of lumps	50	0	40	

(i) Which of the experiments will be finished in less than 3 minutes? **[2]**

(ii) Why is the volume of hydrogen collected, the same in each experiment? **[1]**

(iii) Experiment 1 is repeated with a few drops of copper(II) sulphate added.

The reaction is finished in less than two minutes. Some brown solid is left.

Explain these observations. **[4]**

(Total 10 marks)

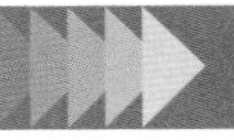

These two answers are at Grades C and A. Compare which one your answer is closest to and think how you could have improved it.

GRADE C ANSWER

This is a balanced equation but it is the wrong equation.

Shreena scores three marks for 3 points on the mark scheme.

Shreena

a $Zn + H_2SO_4 \rightarrow ZnSO_4 + H_2$ ✗✗✗

b (i) 2, 4 and 5 ✓✓

(ii) Same amounts of zinc and acid ✓

(iii) Copper is formed ✓

Copper is a catalyst ✓

Speeds up the reaction ✓

6 marks = Grade C answer

Grade booster ···> move a C to a B

Writing correct chemical equations is an important skill you should practise. It is unfortunate that Shreena has written a correct equation but has used sulphuric acid instead of hydrochloric acid. In b (ii) the use of the word amount is less precise than mass and volume used by Damini.

GRADE A ANSWER

Equation is correct so three marks scored. Damini has included state symbols. These are not needed and are ignored by the examiner.

Damini makes 5 points but only 4 are needed for maximum marks. She could state that the reaction of zinc and copper(II) sulphate is a replacement reaction that takes place because zinc is more reactive than copper (higher in the reactivity series). She could also have included an equation $Zn + CuSO_4 \rightarrow ZnSO_4 + Cu$

This is a better answer than Shreena's.

Damini

a $Zn(s) + 2HCl(aq) \rightarrow ZnCl_2(aq) + H_2(g)$ ✓✓✓

b (i) 2, 4 and 5 ✓✓

(ii) The same mass of zinc and the same volume of acid ✓

(iii) Zinc reacts with copper (II) sulphate ✓

To produce copper ✓

Copper acts as a catalyst ✓

Speeds up the reaction ✓

Brown solid is copper

10 marks = Grade A answer

Grade booster ···> move A to A*

Damini's answer scores full marks. Even she is able to improve this answer by adding additional material. This question is giving her opportunity to impress the examiner. There is nothing in the question aimed at A* candidates.

Try this sample GCSE question and then compare your answers with the Grade C and Grade A model answers on the next page.

In the skeleton Periodic Table five elements are shown. They are shown as V, W, X, Y and Z. These are not the chemical symbols.

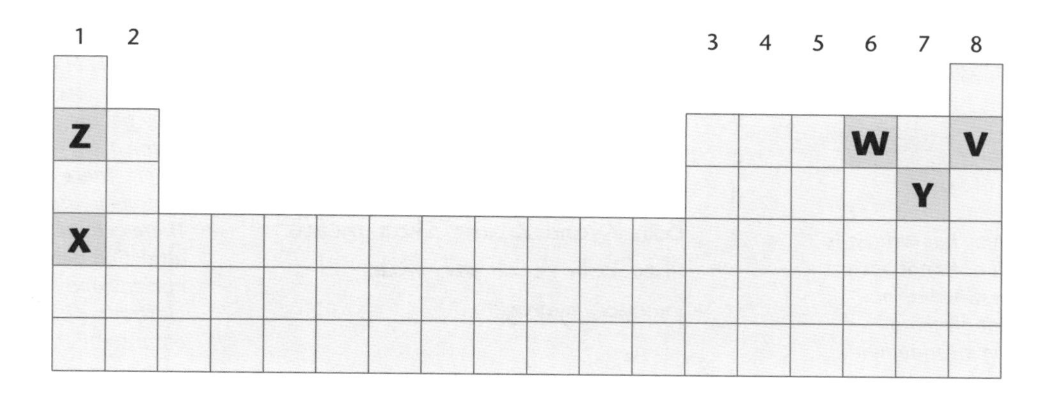

a Which letter represents

(i) a halogen element

(ii) a noble gas? **[2]**

b Which element has atoms with seven electrons in the outer shell? **[1]**

c Write down the electron arrangements of atoms of Z and W. **[2]**

d Explain why atoms of W are smaller than atoms of Z. **[2]**

e Explain why X is more reactive than Z. Use ideas of electron arrangement in your answer. **[3]**

(Total 10 marks)

These two answers are at Grades C and A. Compare which one your answer is closest to and think how you could have improved it.

GRADE C ANSWER

*Amy has answered the first part of the question well, and has gained maximum marks for **a–c**.*

Amy

a (i) Y ✓

(ii) V ✓

b Y ✓

c 2, 1 ✓ 2, 6 ✓

d There are more shells of electrons XX

e Both X and Z are alkali metals. They both react with water to produce hydrogen. XXX

The statement is wrong. Atoms of Z and W both have two shells of electrons. W has five more electrons. They go into the same shell so the atom does not get larger. There are also 5 extra protons in the nucleus. There is increased force of attraction between the nucleus and the other electrons. This makes the atom smaller.

Everything Amy has written is correct but it does not answer the question. She has not taken the hint given and used electron arrangement.

5 marks = Grade C answer

Grade booster ····> move a C to a B

Amy's answer is just Grade C. She falls down with the harder parts d and e. Many candidates do. She should have used ideas of electron arrangement. Her answer to c was correct and should have helped her with d. Look at the information given and the answers you have already made. They can often help with harder parts of the question.

GRADE A ANSWER

This is very good answer. All marks were scored in a–d.

Sanjay

a (i) Y ✓ (ii) V ✓

b Y ✓

c 2,1 ✓ 2,6 ✓

d Atoms of Z and W have the same number of electron shells ✓
Increased nuclear charge attracts outer electrons. ✓

e Both X and Z have one electron in outer shell
Outer electron in X is further away from the nucleus ✓
Weaker force of attraction between outer electron and nucleus ✓

Sanjay has looked at the question, seen it is worth three marks and so has made three points. However, the first point is not on the Examiner's mark scheme.
The answer has just failed to explain that when X and Z react they lose the outer electron, so the outer electron in X is more easily lost.

9 marks = Grade A answer

Grade booster ····> move A to A*

Sanjay has scope to improve his answer in e. This part of the question is targeted at A and A* candidates. He just misses the vital part of the answer. Probably if he had time to re-read his answer he might add to this.

QUESTION BANK

1 The equation for the reaction of sodium thiosulphate solution and dilute hydrochloric acid is:

$$Na_2S_2O_3(aq) + 2HCl(aq) \rightarrow 2NaCl(aq) + H_2O(l) + SO_2(g) + S(s)$$

An experiment was carried out at 20°C to investigate the effect of altering the concentration of sodium thiosulphate on the rate of the reaction.

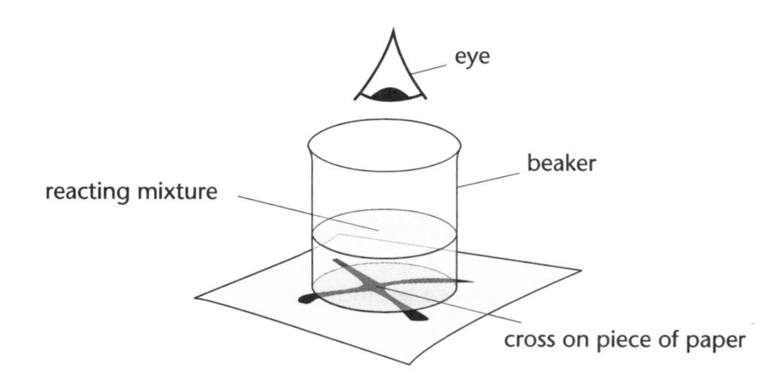

Hydrochloric acid was added to 50 cm^3 of sodium thiosulphate solution. The time was taken until the cross under the beaker disappeared from view.

a) Explain why the cross under the beaker disappears from view.

....

.... ②

b) The results are shown in the table.

Volume of sodium thiosulphate solution in cm^3	Volume of water in cm^3	Volume of hydrochloric acid in cm^3	Time for cross to disappear in s
50	0	5	45
40	10	5	64
30	20	5	82
25	25	5	100
20	30	5	135

i) Suggest three things that were done to make sure that the experiment was a fair test.

1 ....

2 ....

3 .... ③

ii) What can be concluded from these results about the effect of concentration on the rate of reaction?

....

.... ②

c) Describe how this experiment could be modified to examine the effect of temperature on the rate of reaction.

....

....

....

.... (4)

TOTAL 11

---

2 A quick-setting glue is sold in a pack containing two tubes.
Tube A contains a paste.
Tube B contains a catalyst.
When the contents of the two tubes are mixed the glue starts to set.
The setting time is the time taken for the glue to set hard.
The setting time depends upon the amounts of A and B mixed together.
The table shows the setting time when different amounts of B were added to 30 cm^3 of the paste from A.

Volume of B in cm^3	Setting time in s
1	50
3	38
5	26
7	24
9	32
11	40

a) Draw a graph of the volume of catalyst **B** against setting time. (4)

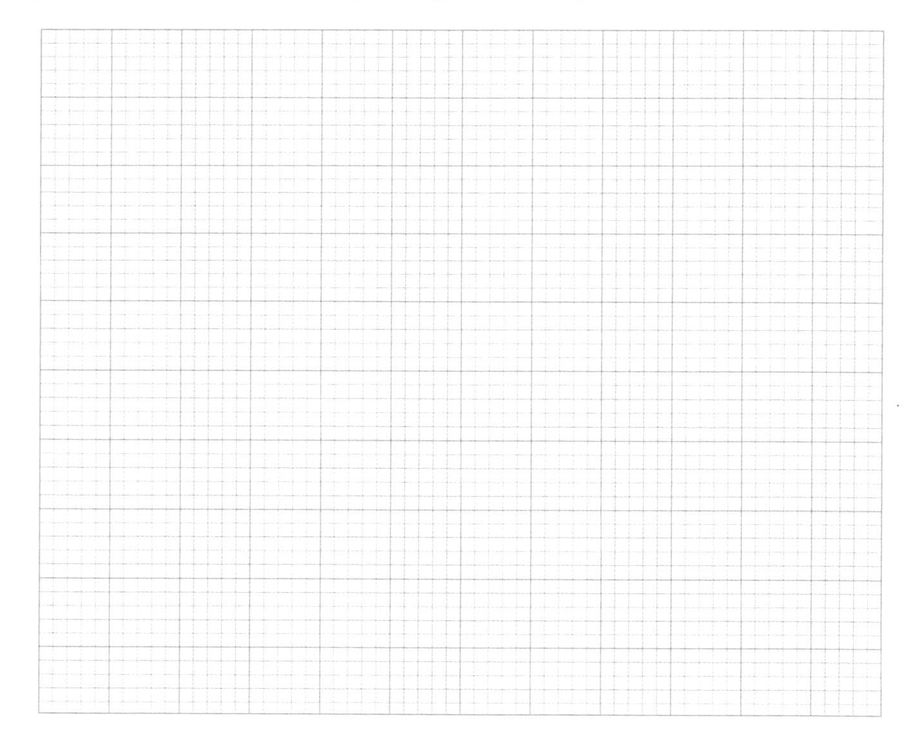

b) i) From the graph, find the shortest setting time.

.... (1)

ANSWERS ON PAGE 71 ANSWERS ON PAGE 71 ANSWERS ON PAGE 71 ANSWERS ON PAGE 71

ii) What percentage of B is present in the mixture with the shortest setting time?
You must show your working.

Answer = ...% ②

TOTAL 7

---

3 The table gives the conditions used for four experiments using zinc and dilute hydrochloric acid.

Experiment	Sample of zinc	Volume of acid in cm^3	Volume of water in cm^3	Temperature in °C	Time in min
1	1.0 g of zinc lumps	50	0	20	3.0
2	1.0 g of zinc powder	50	0	20	2.0
3	1.0 g of zinc lumps	50	50	20	5.0
4	1.0 g of zinc lumps	50	0	40	2.0

Explain the differences in rates of these experiments.
Use your knowledge of particles in your answer. One mark for a clearly ordered answer.

....

....

....

....

....

....

.... ⑥+①

TOTAL 7

4

a) The chart shows the number of electrons in the outer shell for the first eleven elements in the Periodic Table (hydrogen–sodium).

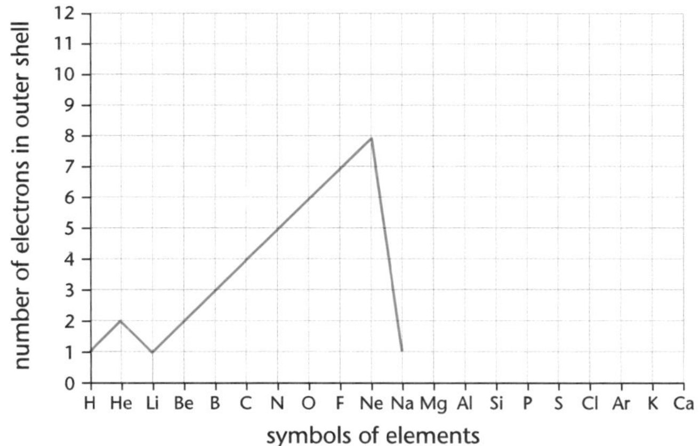

i) Look at the two elements on the two peaks on this graph.
Which family of elements are at the peaks on this graph?

.... ①

ii) Finish the chart for the elements from sodium to calcium. ②

ANSWERS ON PAGE 71 ANSWERS ON PAGE 71 ANSWERS ON PAGE 71 ANSWERS ON PAGE 71

b) The second chart shows the number of electrons in the outer shell for the elements from scandium to krypton.

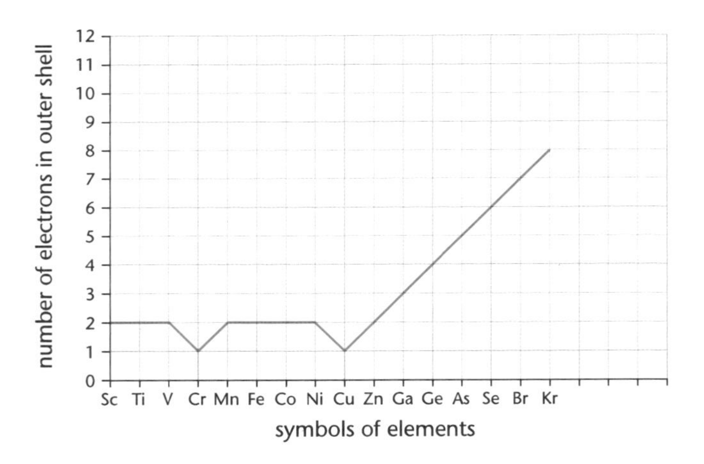

The table gives some information about the transition metals and the alkali metals.

Transition metals

Element	Sc	Ti	V	Cr	Mn	Fe	Co	Ni	Cu	Zn
Atomic number	21	22	23	24	25	26	27	28	29	30
Atomic radii in pm	160	146	131	125	112	123	125	124	128	133

Alkali metals

Element	Li	Na	K	Rb	Cs
Atomic number	3	11	19	37	55
Atomic radii in pm	152	185	231	246	262

i) Cobalt and nickel are both transition metals. Why are the properties of cobalt and nickel similar?

....

.... ②

ii) Explain why the alkali metals increase in reactivity with increasing atomic number.

....

.... ②

iii) Suggest why potassium is more reactive than copper.

....

....

.... ③

TOTAL 10

ANSWERS ON PAGE 71 ANSWERS ON PAGE 71 ANSWERS ON PAGE 71 ANSWERS ON PAGE 71

QUESTION BANK ANSWERS

1 a) Sulphur formed **1**
Sulphur does not dissolve so solution goes cloudy **1**

b)i) Equal volumes of solution **1**
Same temperature **1**
Same volume of hydrochloric acid **1**

Examiner's tip

Answers such as 'use the same apparatus' are incorrect. You are looking for possible variables that are kept constant.

ii) As the concentration of sodium thiosulphate increases the rate of reaction increases **2**

Examiner's tip

One mark could be awarded for statements such as 'it takes less time with more sodium thiosulphate'. This is correct but poorly expressed.

c) Choose one of the experiments **1**
Repeat it four more times heating it to different temperatures **1**
The experiment chosen should not be too fast or less accurate results will be obtained. This suggests probably the 4th or 5th experiment **1**
Candidate suggests between 20 and 60°C or 30, 40, 50, 60 **1**

Examiner's tip

The first two marks are easier to obtain (CC) but the last two marks are at AA. If the temperature is too high too much sulphur dioxide escapes and this is a choking gas. Almost all candidates have done a piece of Sc1 coursework on rates of reaction but few can carry their experiences into written papers.

2 a)

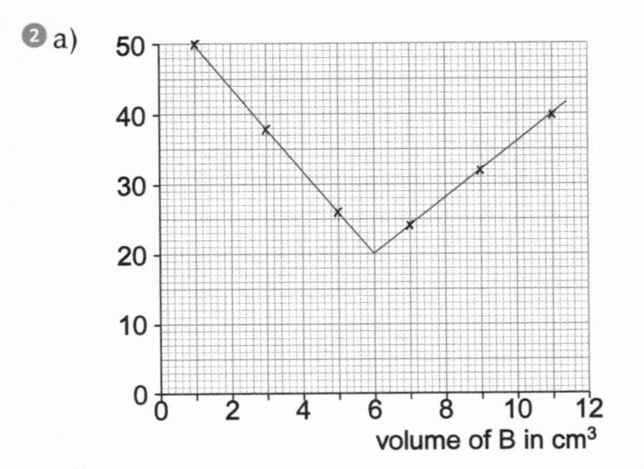

Graph fills over half grid, scales correct **1**
Axes correct and labelled **1**
All plots correct **1**
Two straight lines that intersect **1**

Examiner's tip

On a Higher tier paper you should expect to have to choose scales and axes.
Here you have to draw two lines. In some other questions you might have to deal with anomalous points.

b)i) 20 s (This occurs where the two straight lines cross) **1**

ii) 30 cm³ A and 6 cm³ B

$$\text{Percentage} = \frac{6 \times 100}{36}$$ **1**

$$= 16.7\%$$ **1**

Examiner's tip

Your calculator gives an answer 16.666666. There are too many places of decimals and it should be corrected to 3 significant figures.

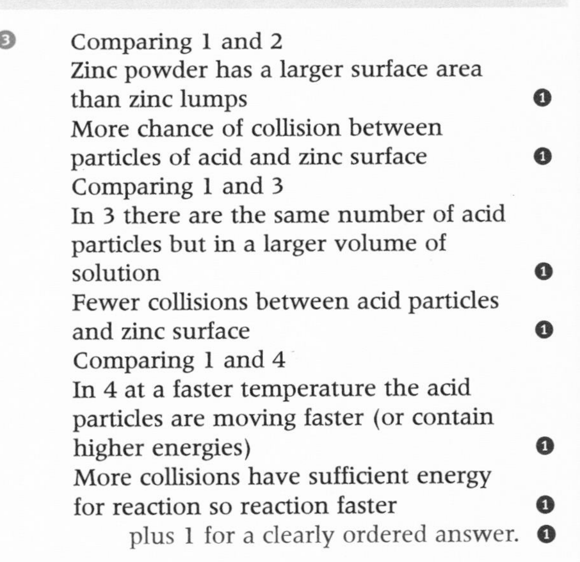

3 Comparing 1 and 2
Zinc powder has a larger surface area than zinc lumps **1**
More chance of collision between particles of acid and zinc surface **1**
Comparing 1 and 3
In 3 there are the same number of acid particles but in a larger volume of solution **1**
Fewer collisions between acid particles and zinc surface **1**
Comparing 1 and 4
In 4 at a faster temperature the acid particles are moving faster (or contain higher energies) **1**
More collisions have sufficient energy for reaction so reaction faster **1**
plus 1 for a clearly ordered answer. **1**

Examiner's tip

There may be different ways of answering a question such as this. However you do this, you must answer in terms of particles.

4 a)i) Noble gases (or group 0) ❶

ii)

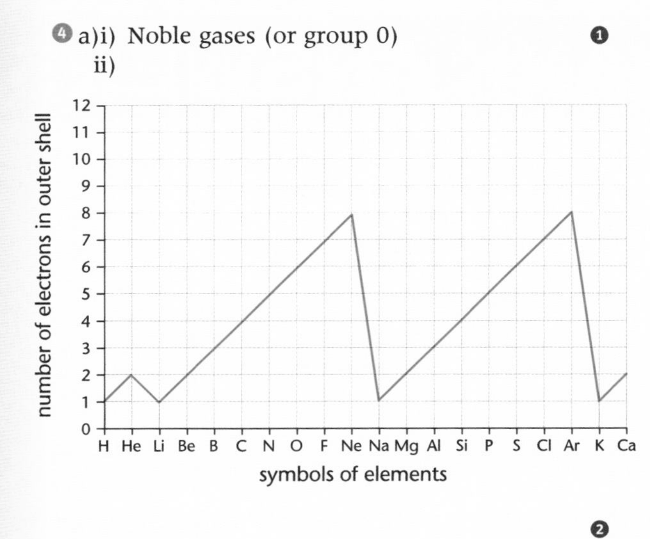

❷

b)i) Cobalt and nickel atoms have the same number of electrons in the outer shell ❶
They are of a similar atomic radius ❶

ii) Atoms increase in size down the group ❶
Force of attraction between the nucleus and outer electron decreases down the group ❶

iii) Both have a single outer electron ❶
Potassium atom is much larger than copper atom ❶
Copper contains more protons so stronger force of attraction on other electron ❶

Examiner's tip

This question requires you to use the data in the question. In looking at reactivity of different elements, look at atomic radius and electron arrangement.

Patterns of behaviour

FOR MORE INFORMATION ON THIS TOPIC ... SEE REVISE GCSE SCIENCE ... CHAPTER 8

CHAPTER 9

Electric circuits

To revise this topic more thoroughly, see Chapter 9 in *Letts Revise GCSE Science Study Guide.*

Try this sample GCSE question then compare your answers with the Grade C and Grade A answers on the next page.

a Electricity from batteries is direct current (d.c.) and that from the mains supply is alternating current (a.c.).

What is the difference between d.c and a.c.? **[1]**

b There are three conductors in the cable that connects a kettle to the mains supply.

(i) When the kettle is operating normally, which conductor has no current passing in it? **[1]**

(ii) Which conductor supplies energy to the kettle? **[1]**

(iii) The current in the kettle element is 9.5 A when it operates from a 240 V supply.

Calculate the power rating of the kettle. **[3]**

The kettle plug is fitted with a fuse rated at 13 A.

(iv) Explain why a fuse rated at 13 A is suitable for the kettle, but one rated at 5 A is not. **[2]**

(v) In which conductor is the fuse placed? **[1]**

(vi) Explain how the fuse acts along with the earth wire to protect users from electric shock. **[3]**

(Total 12 marks)

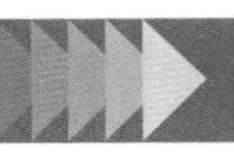

These two answers are at Grades C and A. Compare which one your answer is closest to and think how you could have improved it.

GRADE C ANSWER

Damien

a One comes from batteries and the other is mains ✗

b (i) The earth one (green and yellow) ✓

(ii) The live one ✓

(iii) power = voltage × current ✓
= 240 V × 9.5 A ✓
= 2280 Ω ✗

(iv) The 5 A fuse would blow ✓

(v) In the live ✓

(vi) It stops too much current from wrecking the kettle ✗

No marks are awarded here; the answer merely repeats information given in the question.

This is correct and gains one mark.

There is one mark here for recall of the relationship, one mark for a correct substitution of the physical quantities into the relationship. The final mark is for the correct answer and unit, which this candidate does not gain because the unit is wrong.

Correct – one mark.

This answer gains the mark. Just to be sure, after giving the answer, the candidate has re-enforced it by stating the colour of the insulation. Had it been the wrong colour, no marks would have been gained since there would be two contradictory answers.

This answer gains one mark out of the two available. Only half of the question has been answered.

No marks here for a vague answer which does not describe how the fuse stops an excessive current or how this prevents electric shock.

6 marks = Grade C answer

Grade booster ⋯> move a C to a B

Grade C candidates are often uncertain about the formulae for relationships between physical quantities and the correct units for quantities such as power and resistance. Make sure that you learn all the relationships that are required knowledge and you know the correct units for physical quantities.

GRADE A ANSWER

Kelly

a D.c. is always in the same direction; a.c. changes direction ✓

b (i) The earth wire has no current passing in it. ✓

(ii) Energy is supplied by the live wire ✓

(iii) P = I × V ✓
= 9.5 A × 240 V ✓
= 2280 W ✓

(iv) The 13 A fuse allows the normal current to pass without blowing ✓

(v) In the live ✓

(vi) It blows the fuse if the case becomes live ✓✓

One mark here for a correct description of the difference between a.c and d.c.

Again, one mark for a correct answer.

This answer gives the correct relationship, numerical answer and unit. All three marks are awarded.

One mark for a correct answer.

One mark is awarded for a correct answer.

Only one mark out of two here; the candidate has not explained why the 5 A fuse is not suitable.

Two out of the three marks are gained here. The candidate has the correct cause and effect, but has not explained the mechanism, which is that if the case becomes live a large current passes from live to earth.

10 marks = Grade A answer

Grade booster ⋯> move A to A*

Make sure that you always give full explanations. Look at the number of marks available for each part of a question, and include at least that number of points in your answer. Give full logical explanations for physical effects.

QUESTION BANK

1 When a car is being filled with petrol, the petrol can become negatively charged as it flows down the metal pipe to the petrol tank.

a) Explain how the petrol becomes negatively charged.

....

.... (2)

b) What type of charge does the metal pipe gain?

.... (1)

c) The presence of a person's hand near the metal pipe can cause an explosion. Suggest how this can occur. One mark is for correct spelling, punctuation and grammar.

....

....

.... (3)

d) There is no risk of an explosion if the metal pipe is connected to the bodywork of the car. Explain how this prevents an explosion.

....

.... (2)

TOTAL 8

2

a) The table shows the current in three different electrical appliances when connected to the 240 V mains a.c. supply.

Appliance	Current in A
kettle	9.0
lamp	0.4
toaster	5.5

i) Which appliance has the least resistance? ....

Explain how you can tell from the data in the table.

....

.... (2)

There are three wires in the lead that connects the toaster to the mains supply.

ii) Two of these are the live and the earth. Write down the name of the third wire.

.... (1)

The lamp needs only two wires to connect it to the mains supply.

iii) Which connection is not needed? Explain why it is not needed.

....

....

.... (3)

b) Calculate the power rating of the toaster when it is operating from the 240 V mains supply.

....

....

.... (3)

TOTAL 9

ANSWERS ON PAGE 79 ANSWERS ON PAGE 79 ANSWERS ON PAGE 79 ANSWERS ON PAGE 79

3 Electricity retailers sell energy to domestic customers in units of kWh.
The cost of each kWh is 7p.
The cost of using an appliance is calculated using the relationship:

$$\text{cost} = \text{power in kW} \times \text{time in h} \times \text{cost of 1 kWh}$$

In a household, a 2.5 kW immersion heater is in use for 1.5 hours each day.

a) Calculate the cost of the energy transferred to the heater in a period of 90 days.

....

.... (2)

b) The heater is accidentally left switched on overnight, for a period of 12 hours.
Explain why the energy transferred to the heater in this time is much less than
$12 \times 2.5 = 30$ kWh.

....

.... (2)

c) There are three conductors in the cable to the heater. These are the live, earth and neutral wires.

i) Describe the functions of the live and neutral conductors.

....

.... (2)

ii) Explain how the earth conductor, together with the fuse, protects the user from electrocution.

....

....

.... (3)

TOTAL 9

---

4 The diagram shows a circuit used to investigate the resistance of a carbon rod.

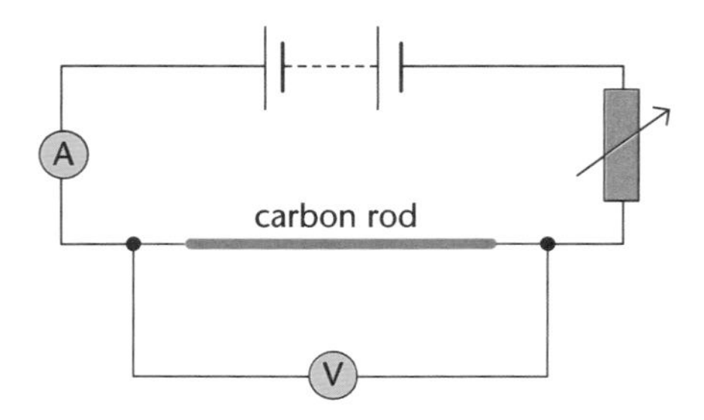

a) Write down **two** ways of changing the current in the circuit.

....

.... (2)

b) The table shows the ammeter and voltmeter readings for a range of current passing in the rod.

voltage in V	current in A
0.8	0.4
1.9	1.1
3.1	2.1
4.1	3.2
4.7	4.1

ANSWERS ON PAGE 79 ANSWERS ON PAGE 79 ANSWERS ON PAGE 79 ANSWERS ON PAGE 79

i) Use the grid to plot a graph of voltage against current. ②

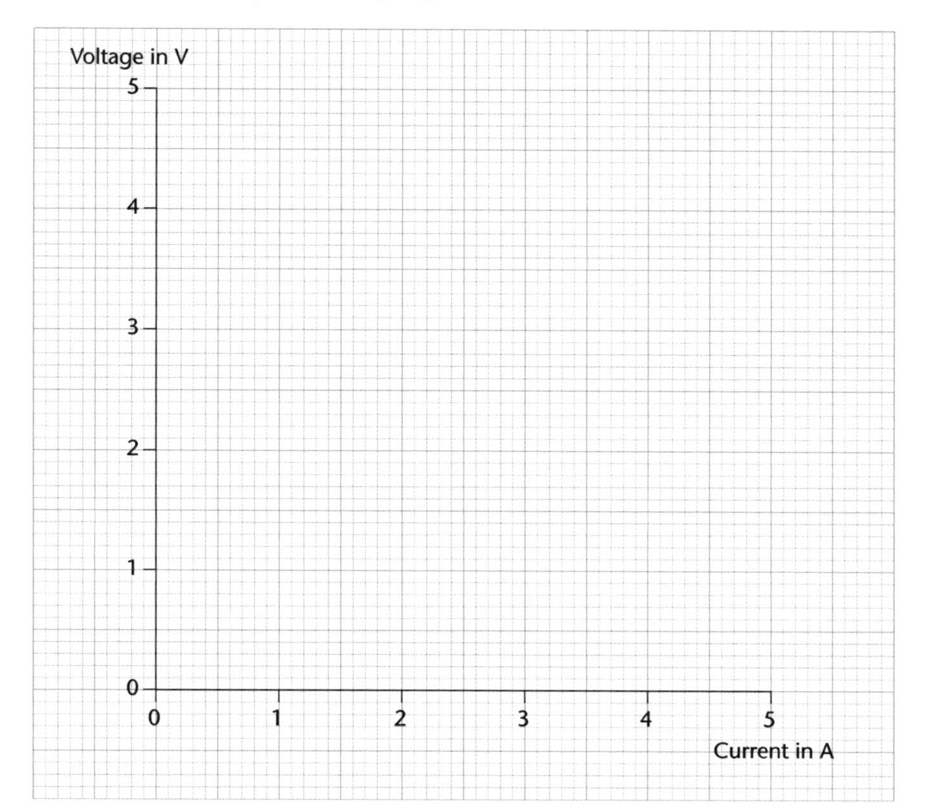

ii) Read from your graph the value of the voltage when the current in the carbon rod is 1.5 A.

.... ①

iii) Calculate the resistance of the carbon rod when the current in it is 1.5 A.

....

....

.... ③

TOTAL 8

---

5 An electric blanket has two heating elements. These can be connected either in series or in parallel to the mains supply. The diagram shows the two different ways of connecting the elements to the mains supply.

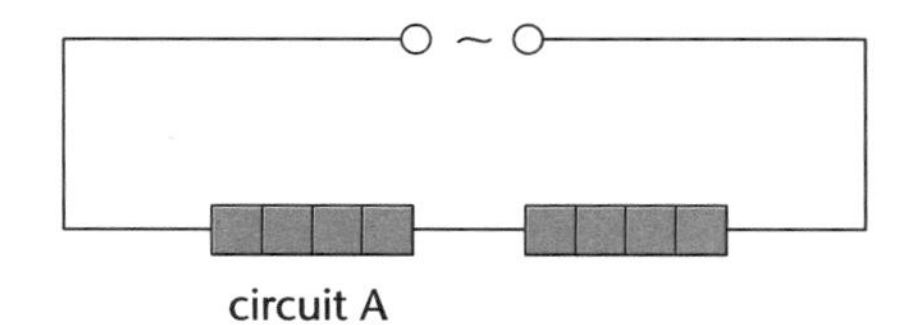

circuit A

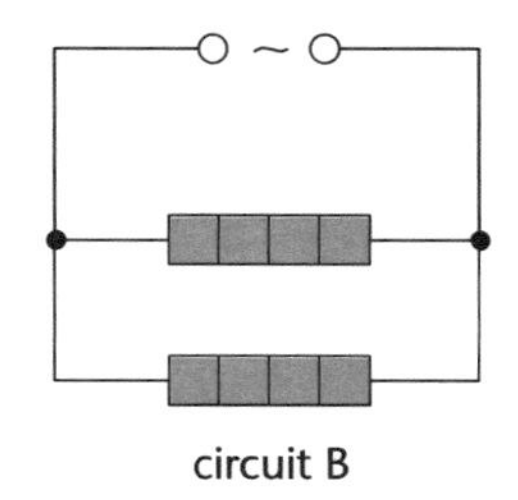

circuit B

a) Which circuit has the greater current?
Explain why this is.

....

.... ②

b) The mains voltage is 240 V.
Each heating element has a resistance of 480 Ω.

ANSWERS ON PAGE 79 ANSWERS ON PAGE 79 ANSWERS ON PAGE 79 ANSWERS ON PAGE 79

i) Calculate the current passing in circuit A.

....

....

.... ③

ii) Calculate the total power of both heaters in circuit A.

....

....

.... ③

c) Explain whether the power output of the heaters in circuit B is greater or smaller than that of the heaters in circuit A.

....

....

.... ③

TOTAL 11

---

6 The diagram represents the electrolysis of a salt solution.

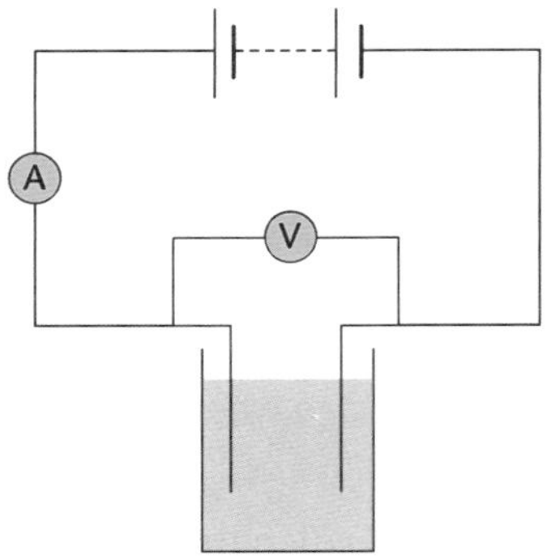

A
V

During electrolysis current passes in both the copper wires and in the salt solution.

a) Describe the difference in the mechanism of conduction in the wire and the solution.

....

....

....

....

.... ④

b) The current in the solution is 0.80 A.
Calculate the quantity of charge that flows through the solution in 35 s.

....

....

.... ③

c) The voltmeter reading is 4.5 V.
How much energy is transferred to the solution in 35 s?

....

....

.... ③

TOTAL 10

ANSWERS ON PAGE 79 ANSWERS ON PAGE 79 ANSWERS ON PAGE 79 ANSWERS ON PAGE 79

QUESTION BANK ANSWERS

1 a) The petrol gains electrons 1
From the metal pipe 1
b) Positive 1

Examiner's tip

All electrostatic charging is due to the transfer of electrons between objects. The outermost electrons in atoms and molecules are easily removed by the friction forces that act when objects rub together. A common error by GCSE candidates is to state that objects become positively charged by gaining protons.

c) The metal pipe becomes charged to a high voltage 1
This ionises the air between the pipe and the person's hand, causing a spark 1
Quality of written communication – this mark is awarded to an answer written in complete sentences with correct spelling and use of full stops and capital letters to end and begin sentences. 1
d) The charge becomes spread out 1
So a high voltage does not occur 1

Examiner's tip

High voltages can be caused by small amounts of charge if the charge is concentrated over a small area. Spreading the charge out over a larger area reduces the voltage.

2 a)i) The kettle 1
The kettle has the greatest current 1

Examiner's tip

If you calculated the resistance of the kettle, you also need to calculate the resistance of the lamp and the toaster to make a comparison.

ii) Neutral 1
iii) The earth is not needed 1
The lamp has no exposed metal parts 1
So a user cannot get a shock by touching the lamp 1
b) $P = I \times V$ 1
$= 5.5\,\text{A} \times 240\,\text{V}$ 1
$= 1320\,\text{W}$ 1

Examiner's tip

A correct answer and unit to a numerical question always gains full marks. However, a wrong answer gains no marks so you should always show each step in the calculation. This way, you can gain two marks out of three for the wrong answer. You should note that, to gain the final mark, both the numerical answer and the unit must be correct.

3 a) cost = $2.5\,\text{kW} \times 90 \times 1.5\,\text{h} \times 7\,\text{p/kWh}$ 1
= 2362.5p or £23.625 1

Examiner's tip

When needed, this relationship will be given in the question or on a separate formula sheet. It is quite acceptable to leave the answer in pence, but if you do convert it into pounds, make sure that you do it correctly! A wrong conversion could cost you a mark.

b) The heater is fitted with a thermostat to turn it off when the water reaches a certain temperature 1
So it is not heating the water for 12 hours 1

Examiner's tip

The thermostat is an additional switch in series with the main switch; it turns the heater off when the pre-set temperature has been reached and on again when the water cools to a few degrees below this temperature.

c)i) The live conductor carries energy to the heater 1
The neutral conductor completes the circuit 1
ii) The earth conductor is connected to the metal casing 1
If this becomes live there is a low-resistance path to earth 1
The resulting high current melts the fuse and breaks the circuit 1

Examiner's tip

Always try to use the correct technical and scientific terms. In this case it is correct to state that the fuse 'melts' rather than 'blows'.

Electric circuits

4 a) Change the voltage
Change the length or diameter of the carbon rod
Adjust the setting of the variable resistor
any two, 1 mark each 2

Examiner's tip

A common error on this type of question is to answer 'use a bigger battery or power supply'. This does not gain a mark as it is the voltage, rather than the physical size, that determines the current in the circuit.

b)i) Your completed graph should look like this:

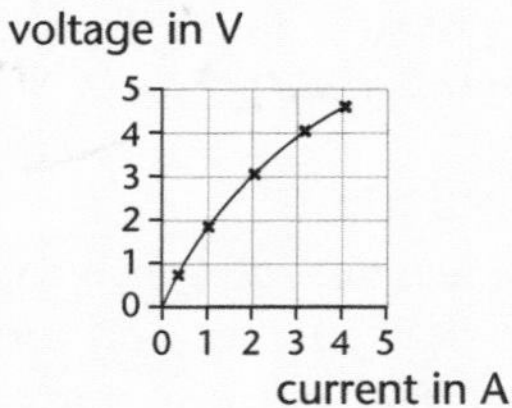

Award marks for:
Correct plotting of all the points 1
Drawing a smooth curve 1

Examiner's tip

You do not gain the last mark if you joined the points together with a series of straight lines.

ii) 2.4 V 1

Examiner's tip

When marking answers from candidates' reading of their graphs, examiners usually allow some leeway; in this case any answer in the range 2.3 V to 2.5 V would be marked as correct.

iii) $R = V \div I$ 1
$= 2.4\ V \div 1.5\ A$ 1
$= 1.6\ \Omega$ 1

Examiner's tip

Full marks are allowed for a correct calculation using your answer to ii) for the voltage if your answer was not 2.4 V.

5 a) Circuit B 1
There is less resistance/there are two current paths 1

b)i) $I = V \div R$ 1
$= 240\ V \div 960\ \Omega$ 1
$= 0.25\ A$ 1

Examiner's tip

When resistors are connected in series, the total resistance is the sum of the individual resistances.

ii) $P = I \times V$ 1
$= 0.25\ A \times 240\ V$ 1
$= 60\ W$ 1

Examiner's tip

If your answer to b) i) was wrong, you still gain full marks here provided that you have carried out the calculation correctly.

c) Greater, as there is less resistance in the circuit 1
So the current is greater 1
And each heater operates from the full mains voltage 1

6 a) Current is a flow of electric charge 1
In the metal the charge carriers are negatively-charged electrons 1
In the salt solution the charge flow is due to positively-charged ions moving in one direction 1
And negatively-charged ions moving in the opposite direction 1

Examiner's tip

In metals only the free electrons can move, but in molten or dissolved electrolytes and in ionised gases both the positive and the negative ions move to form a current.

b) $Q = I \times t$ 1
$= 0.80\ A \times 35\ s$ 1
$= 28\ C$ 1

c) energy transfer = voltage × charge
or $E = V \times Q$ 1
$= 4.5\ V \times 28\ C$ 1
$= 126\ J$ 1

Examiner's tip

The voltage across a power supply or a component is the energy transfer for each coulomb of charge that flows through it. In the case of a power supply the energy is transferred to the charge, but as it flows through a component the energy is transferred from the charge.

CHAPTER 10

Force and motion

To revise this topic more thoroughly, see Chapter 10 in *Letts Revise GCSE Science Study Guide.*

Try this sample GCSE question and then compare your answers with the Grade C and Grade A model answers on the next page.

The graph shows how the velocity of a car changes when the driver leaves home to go to work.

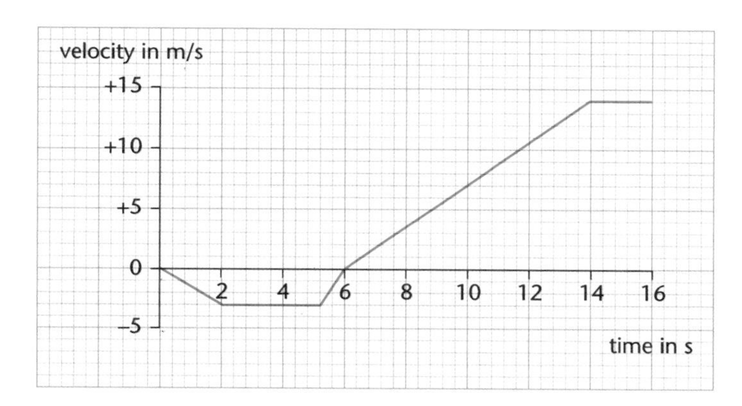

A positive velocity represents a forwards movement.

a **(i)** In the 16 s shown on the graph, for how long was the car reversing? **[1]**

(ii) In the 16 s shown on the graph, for how long was the car travelling at a constant speed? **[1]**

b **(i)** Calculate the acceleration of the car during the first 2 s shown on the graph. **[3]**

(ii) The total mass of the car and its driver is 900 kg.
Calculate the size of the resultant force on the car during this acceleration. **[3]**

c **(i)** What was the greatest forwards speed of the car? **[1]**

(ii) Calculate the distance travelled by the car between the times of 6 s and 14 s shown on the graph. **[3]**

(Total 12 marks)

These two answers are at Grades C and A. Compare which one your answer is closest to and think how you could have improved it.

GRADE C ANSWER

Peter

a (i) 6 seconds ✓

(ii) 1.3 s ✗

b (i) acceleration = change in speed ÷ time = 3 ÷ 2 = 1.5 m/s ✓✓

(ii) Force = mass × Earth's pull = 9000 kg ✗

c (i) 14 m/s ✓

(ii) speed × time = 14 × 14 = 196 m. ✓✗✗

One mark is awarded here for identifying the part of the graph that shows the car reversing and reading the time scale correctly.

Peter does not gain the mark here. He has made a number of errors. He has not appreciated that the car spends some time travelling at a constant speed in each direction and he has only attempted to give the time spent travelling at a constant speed when the car was reversing.
He has mis-read the scale on the graph. Always double-check so that you are sure what each scale division represents.

This answer is correct apart from the wrong unit – this is a common error by GCSE candidates. He gains two marks out of three.

No marks are awarded for Peter's response to this part of the question. He has tried to work out the size of the Earth's downward pull on the car, rather than that of the forwards push. Once again, he has the wrong unit.

Correct – one mark.

This answer gains one mark out of the three. He has made two errors: He does not realise that when the speed of an object is not constant he should use the average speed when working out the distance travelled.
He has read the time scale for the end of the accelerated motion correctly, but not realised that this did not start until the car had been moving for 6 s, so the time accelerating forwards is only 8 s, not 14 s.

5 marks = Grade C answer

Grade booster ⋯> move a C to a B

Grade C candidates often omit units or do not know the units for key quantities such as acceleration. Make sure that you know the correct units for each physical quantity that you come across in your GCSE course.

GRADE A ANSWER

Siobhan

a (i) 6.0 s ✓

(ii) 3.2 s + 2.0 s = 5.2 s ✓

b (i) acceleration = increase in velocity ÷ time taken ✓ = 3.0 m/s ÷ 2.0 s = 1.5 m/s² ✓✓

(ii) force = mass × acceleration ✓ = 900 × 1.5 = 1350 kg ✓ ✗

c (i) 14 m/s ✓

(ii) distance = speed × time ✓ = 14 m/s × 8 s = 112 m. ✓ ✗

Correct – one mark.

Correct – Siobhan has read the scales correctly and realised that there are two separate time intervals when the car was travelling at constant speed.

Full marks here – notice how Siobhan has always written down the unit with each physical quantity – this is good practice.

This time she omitted the units when she substituted the physical quantities into the relationship – and ended up with the wrong unit for force. Two marks out of three for an answer that is numerically correct.

This is correct and gains the mark.

Siobhan gains two marks out of three here; she has not appreciated the significance of using the average speed, so her answer is twice the correct one.

10 marks = Grade A answer

Grade booster ⋯> move A to A*

An A* candidate should never make an error with units and should show that she/he appreciates and understands why *distance travelled = average speed × time* rather than just *speed × time*.

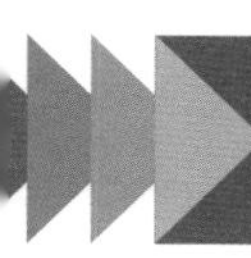

QUESTION BANK

1 The graph shows how the speed of a train changes as it pulls away from a station.

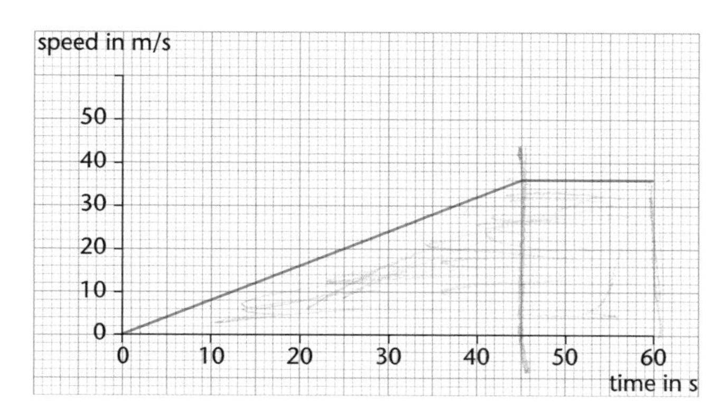

a) Without doing any calculations, describe the motion that this shows.

....

.... ②

b) Calculate the acceleration of the train:

i) during the first forty-five seconds

....

....

.... ③

ii) during the next fifteen seconds.

.... ①

c) Calculate the total distance travelled by the train in the first 60 s after pulling away from the station.

....

....

.... ④

TOTAL 10

---

2 The data in the table shows how the thinking and braking distance of a car depends on its speed.

Speed in m/s	Thinking distance in m	Braking distance in m	Stopping distance in m
5.0	3.0	1.5	
10.0	6.0	6.0	
15.0	9.0	13.5	
20.0	12.0	24.0	
25.0	15.0	37.5	

a) Describe what is meant by:

i) 'thinking distance'

.... ①

ANSWERS ON PAGE 87 ANSWERS ON PAGE 87 ANSWERS ON PAGE 87 ANSWERS ON PAGE 87

ii) 'braking distance'.

.... ①

b) i) Compare the thinking times at speeds of 10 m/s and 20 m/s.

....

....

.... ④

ii) Explain the result of your comparison in i).

....

.... ②

c) i) Compare the braking times at speeds of 10 m/s and 20 m/s.

....

....

.... ②

ii) Explain the result of your comparison in i).

....

.... ②

d) When the speed of a vehicle doubles, the thinking distance also doubles but the braking distance is multiplied by four.
Explain why.

....

....

.... ③

e) i) Complete the table by calculating the stopping distances. ①

The maximum allowed speed of a vehicle travelling in an area where there are houses and schools is 13 m/s.

ii) Estimate the stopping distance of a vehicle travelling at this speed.

.... ①

iii) Explain the dangers posed by drivers who exceed this maximum.

....

....

....

.... ④

TOTAL 21

---

3 A skydiver jumps out of an aircraft and falls towards the Earth.

a) What two forces act on the skydiver? State the direction of each force.

....

.... ④

The graph shows how the speed of the skydiver changes before the parachute is opened.

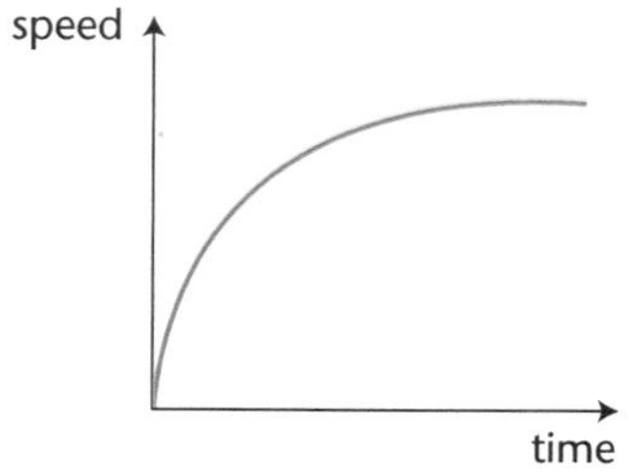

speed

time

ANSWERS ON PAGE 87 ANSWERS ON PAGE 87 ANSWERS ON PAGE 87 ANSWERS ON PAGE 87

b) Which part of the graph shows the skydiver travelling at terminal velocity?

.... (1)

c) Use the graph to describe how the acceleration of the skydiver changes.

....

.... (2)

d) In terms of the forces acting on the skydiver, explain why the acceleration changes in this way.

....

....

.... (3)

TOTAL 10

---

4 A motorist is driving along a road and sees a traffic light at red. The graph shows the speed of the car from the instant that the motorist notices the traffic light.

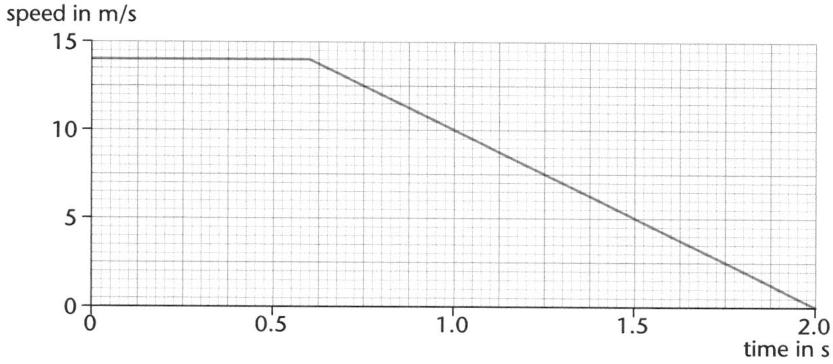

a) What was the speed of the car when the driver noticed the traffic light?

.... (1)

b) How long did it take the driver to react to the traffic light?

.... (1)

c) State **two** factors that could affect the driver's reaction time.

....

.... (2)

d) Calculate the distance travelled by the car before the driver reacted to the traffic light.

....

.... (3)

e) What feature of the graph represents:

i) the distance travelled while the car was braking

....

.... (2)

ii) the deceleration of the car during braking?

.... (1)

f) Calculate the deceleration of the car during braking.

....

....

.... (4)

TOTAL 14

ANSWERS ON PAGE 87 ANSWERS ON PAGE 87 ANSWERS ON PAGE 87 ANSWERS ON PAGE 87

5 The diagram shows the forces acting on a car.

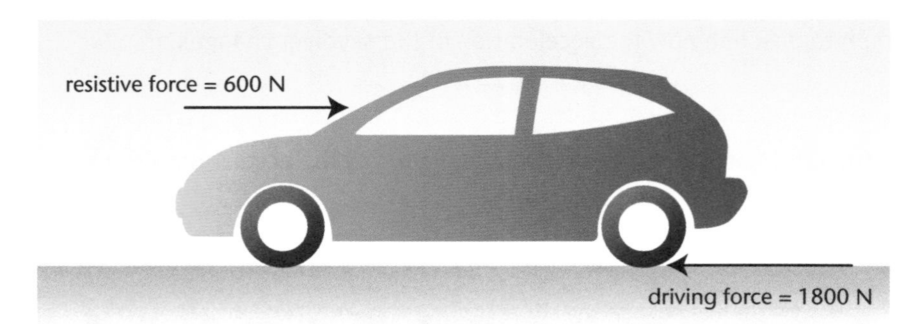

a) How can you tell from the diagram that the car is accelerating in the forwards direction?

.... (1)

b) Calculate the size of the unbalanced force on the car and state its direction.

....

.... (2)

c) The total mass of the car and its driver is 900 kg.
Calculate the acceleration of the car.

....

....

.... (3)

d) Explain how the acceleration of the car is affected when the car is fully loaded for a family holiday. One mark is for correct spelling, punctuation and grammar.

....

....

.... (3)

e) Describe how the balance of the forces acting on the car changes as it accelerates and then maintains a constant speed.

....

....

.... (3)

TOTAL 12

ANSWERS ON PAGE 87 ANSWERS ON PAGE 87 ANSWERS ON PAGE 87 ANSWERS ON PAGE 87

QUESTION BANK ANSWERS

1 a) The train accelerates for 45 s (1)
It then travels at a constant speed (1)

Examiner's tip
Take care to be precise with your answers. The term 'steady speed' is often used by examination candidates but it is too vague to be given credit.

b)i) Acceleration = increase in velocity ÷ time taken (1)
= 36 m/s ÷ 45 s (1)
= $0.80\ m/s^2$ (1)

Examiner's tip
In this case there is no change in direction, so the increase in velocity is equal to the increase in speed.

ii) Zero (1)

Examiner's tip
The gradient of a velocity–time or speed–time graph represents acceleration. After 45 s the gradient of the graph is zero, showing that the train is not accelerating.

c) Distance travelled during the first 45 s = average speed × time (1)
= $\frac{1}{2}$ × 36 m/s × 45 s (1)
= 810 m (1)
Distance travelled during next 15 s
= 45 m/s × 15 s = 675 m
Total distance travelled = 810 m + 675 m = 1485 m (1)

Examiner's tip
On a speed–time or a velocity–time graph, the distance travelled is represented by the area between the graph line and the time axis.

2 a) i) The distance that the car travels during the driver's reaction time or thinking time. (1)

Examiner's tip
The driver's reaction time is increased by the effects of drugs or alcohol or if the driver is talking on a mobile phone or paying attention to the car's hi-fi system.

ii) The distance the car travels during braking (1)

Examiner's tip
Braking distance depends on the weather and road surface conditions, and of course on the conditions of the car's brakes and tyres.

b)i) At 10 m/s: thinking time
= thinking distance ÷ speed (1)
= 6.0 m ÷ 10 m/s (1)
= 0.60 s (1)
At 20 m/s: thinking time = 12.0 m ÷ 20 m/s = 0.60 s (1)
ii) The thinking time depends on the driver's state and alertness (1)
It is independent of the speed of the car (1)

Examiner's tip
If thinking time is fixed, thinking distance is directly proportional to the speed of the car.

c)i) At 10 m/s: braking time = braking distance ÷ speed = 6.0m ÷ 10.0s = 0.6s (1)
At 20 m/s: braking time = 24.0 m ÷ 20 m/s = 1.20 s
ii) The car has to brake from twice the speed (1)
For the same deceleration, this takes twice the time (1)

d) The thinking distance doubles because the car travels double the distance in the same time interval (1)
The braking distance is multiplied by four because the braking time is doubled (1)
And the average speed during braking is also doubled (1)

Examiner's tip
Each of these two factors is doubled, so the overall effect is a multiplication by four.

e)i) The completed last column of the table is: (1)

Stopping distance in m
4.5
12.0
22.5
36.0
52.5

ii) Estimate in the range 17 m to 20 m. (1)
iii) The thinking and braking distances are both increased (1)
So if a hazard occurs the car needs a longer stopping distance (1)
This increases the risk of the car hitting a hazard such as a person stepping into the road (1)
And also means that in any collision the force would be greater (1)

Examiner's tip
The force would be greater because the car would collide at a higher speed.

3 a) The sky diver's weight or the Earth's pull (1)
acts downwards (1)
Air resistance (1)
acts upwards (1)

Examiner's tip

Avoid using the term 'gravity' when describing forces. The force between the skydiver and the Earth is a gravitational force because it is due to their masses, but the Earth is the actual object that pulls on the skydiver.

b) The last part, where the curve is parallel to the time axis (1)

c) The acceleration is decreasing (1)
And is zero towards the end of the graph (1)

Examiner's tip

The gradient of the graph decreases to zero, representing an acceleration that also decreases to zero; i.e. the skydiver travels at a constant speed.

d) The downward force (weight) stays constant but the upward force increases as the sky diver travels faster. (1)
As the upward force increases, the size of the unbalanced force, and therefore the acceleration, decreases. (1)
When the upward and downward forces are equal in size the unbalanced force on the skydiver is zero, so there is no acceleration. (1)

4 a) 14 m/s (1)

Examiner's tip

This mark is for data analysis, it is testing whether you can read graph scales correctly.

b) 0.6 s (1)

c) Drugs; alcohol; loud music or children; tiredness any two (2)

d) distance travelled = speed × time (1)
= 14 m/s × 0.6 s (1)
= 8.4 m (1)

Examiner's tip

Showing the intermediate steps in a calculation allows you to gain some credit even if the final answer is wrong.

e) i) The area between the graph line and the time axis (1)
Between 0.6 s and 2.0 s (the sloping part of the line) (1)

Examiner's tip

If you answered 'the area between the graph line and the time axis', you have included the distance travelled during the driver's reaction time. Always make your answers are as precise as possible so that there can be no doubt in the mind of the examiner. Answers that could have more than one meaning are marked wrong!

ii) The gradient of the graph between 0.6 s and 2.0 s (the sloping part of the line) (1)

Examiner's tip

Deceleration is an alternative name for a negative acceleration, used to describe a situation where a vehicle or other object is slowing down.

f) deceleration = change in velocity ÷ time taken (1)
= 14 m/s ÷ 1.4 s (2)
= 10 m/s^2 (1)

Examiner's tip

The two marks in line two are for correct readings of speed and time from the graph and correct substitution into the equation. If your graph readings were wrong but you substituted the wrong values correctly, only 1 mark is awarded. Even if your readings were wrong, you still gain the last mark for working out the answer and giving the correct unit.

5 a) The forwards-directed force (the driving force) is greater than the backwards-directed force (the resistive force) (1)

Examiner's tip

When the forces on an object are unbalanced, it accelerates in the direction of the unbalanced force.

b) 1200 N (1)
Forwards (1)

Examiner's tip

Forces acting in the same direction add together, but the effect of those acting in opposite directions is the difference between them.

c) acceleration = force ÷ mass (1)
= 1200 N ÷ 900 kg (1)
= 1.33 m/s^2 (1)

d) The mass of the car is increased (1)
This reduces the acceleration for the same driving force (1)
Quality of written communication – answer written in complete sentences with correct punctuation and grammar. (1)

e) At low speeds, the resistive forces acting on the car are small (1)
The size of the resistive forces increases with increasing speed (1)
When the car travels at a constant speed the resistive forces and the driving force are balanced (1)

Examiner's tip

If a vehicle is not changing its speed or direction then the forces acting on it are balanced.

FOR MORE INFORMATION ON THIS TOPIC ... SEE REVISE GCSE SCIENCE ... CHAPTER 10

CHAPTER 11

Waves

To revise this topic more thoroughly, see Chapter 11 in *Letts Revise GCSE Science Study Guide.*

Try this sample GCSE question and then compare your answers with the Grade C and A model answers on the next page.

A television set used at home is designed to detect waves with a wavelength of around 0.60 m.
The speed of the waves through the air is 3.00×10^8 m/s.

a Calculate the frequency of a wave that has a wavelength of 0.60 m. **[3]**

The waves that carry national television signals are sent from the Telecom tower to a repeater station. Here they are amplified before being sent to other repeater stations. Dish aerials are used to receive and transmit the waves.

The diagram shows what happens when waves with a wavelength of 0.60 m are transmitted from a dish aerial that has a diameter of 1.50 m.

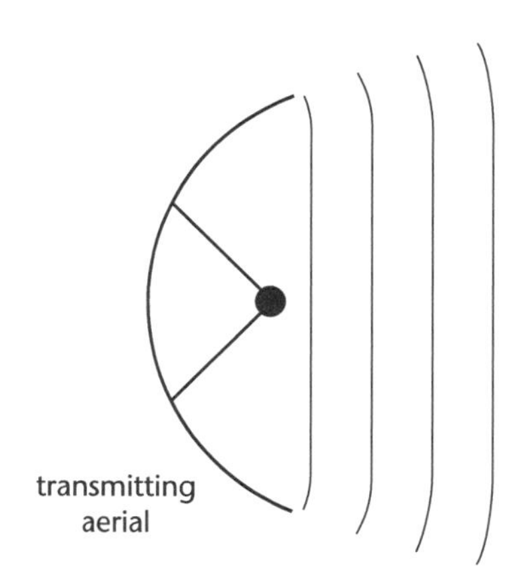

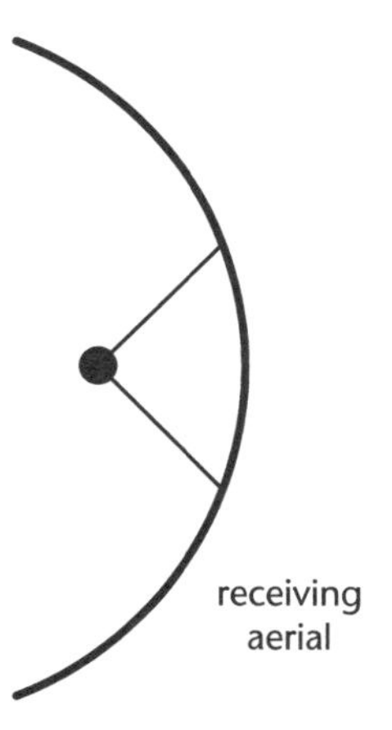

b (i) What happens to the waves as they leave the dish? **[1]**
(ii) Explain why this is undesirable. **[2]**
(iii) How could changes to the dish reduce this effect? **[1]**
(iv) Explain why this is impractical. **[2]**

c The signals transmitted from the Telecom tower are carried on waves that have a wavelength of 2.5 cm.
Suggest why this is a suitable wavelength to use. **[3]**

(Total 12 marks)

GRADE C ANSWER

Petra

The relationship and substitution are correct, as is the final numerical answer, but the unit is wrong – it should be hertz (Hz). Two marks out of three are awarded.

a frequency = speed ÷ wavelength ✓
= 3.00×10^8 m/s ÷ 0.60 m
= 5.00×10^8 s ✓ ✗

b (i) The waves become curved round the edges; it is refracted. ✗

No marks here for confusing refraction and diffraction.

(ii) The receiving aerial needs to detect straight waves. ✓ ✗

True, and it will. But Petra has not realised that the spreading out of the transmitted waves results in a weak signal being received.

(iii) By making it less bent. ✗

No – this would not work. The size of the dish needs to be changed. No marks.

(iv) The metal is hard to unbend. ✗

Petra is completely on the wrong track. Again, no marks.

c This is a much shorter wavelength and will not curve round as much. ✗

The first statement is correct, but not enough to gain any credit. A correct response would be to compare the wavelength with the size of the dish used.

3 marks = Grade C answer

Grade booster ····> move a C to a B

This is a very difficult question for grade C candidates, and so it is not surprising that Petra has struggled to glean a few marks. Diffraction is a difficult topic, and is not one that you would be able to do more than recognise on foundation tier papers. However, at higher tier you need to be able to link the effects of diffraction to the wavelength and the size of the aperture where diffraction takes place.

GRADE A ANSWER

Tamsin

Three marks out of three for a fully correct answer.

a frequency = speed ÷ wavelength ✓
= 3.00×10^8 m/s ÷ 0.60 m ✓
= 5.00×10^8 Hz ✓

b (i) They are diffracted. ✓

Correct – one mark.

(ii) The waves spread out from the transmitting aerial. ✓

Correct for one mark, but Tamsin has not gone on to state that this results in a weak signal being received at the receiving aerial – so only one mark out of two.

(iii) By increasing the diameter of the aerial ✓

Correct again – one mark.

(iv) The aerial would be too big. ✓

This gains one mark out of two. Tamsin has not explained why the aerial would be too big (it would need to be bigger than the tower could support) or given an estimate of the size of the aerial required.

c This wavelength is much smaller than the diameter of the dish so less spreading of the waves would occur. ✓✓

This answer is correct but does not go quite far enough. Tamsin has made two valid points and gains two marks. For the third mark she would need to state that the dish diameter is many wavelengths wide for this wavelength.

9 marks = Grade A answer

Grade booster ····> move A to A*

A* candidates need to be fully conversant with difficult topics such as diffraction. They need to give fully reasoned answers and sufficient explanation to convey that they understand the topic fully. An A* candidate should be able to give a reasoned argument about how the amount of spreading of a wave depends on the size of the wavelength relative to the aperture.

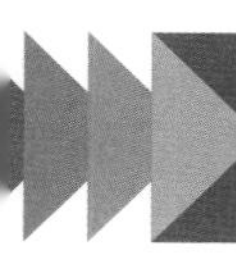

QUESTION BANK

1

a) The diagram shows how water waves spread out after passing through a gap.

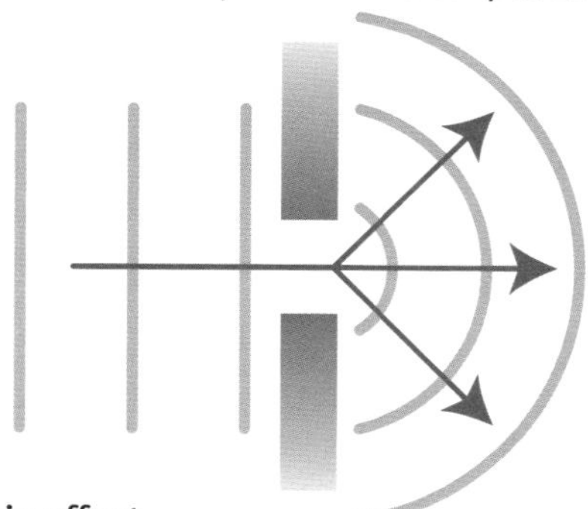

i) Write down the name of this effect.

.... (1)

ii) What two factors determine the amount of spreading that takes place?

1 ....

2 .... (2)

iii) A typical sound wave has a wavelength of 1 m.
A typical light wave has a wavelength of 5.0×10^{-7} m.
Explain why sound spreads out after passing through a doorway but light does not.

....

....

.... (2)

b) A loudspeaker system used in a hi-fi set consists of two loudspeakers; one has a diameter of 6 cm and one a diameter of 30 cm.

The loudspeaker reproduces sounds that range in frequency between 20 Hz and 15 000 Hz.

i) The speed of sound in air is 330 m/s.
Calculate the wavelength in air of a sound that has a frequency of 15 000 Hz.

....

....

.... (3)

ii) Explain why a sound of this wavelength should be reproduced using the smaller loudspeaker.

....

.... (2)

TOTAL 10

ANSWERS ON PAGE 95 ANSWERS ON PAGE 95 ANSWERS ON PAGE 95 ANSWERS ON PAGE 95

2 X-rays and ultrasound are both used in medicine to create images of the inside of the body.

a) X-rays have a similar range of wavelengths and frequencies to that of gamma rays. What is the difference between X-rays and gamma rays?

....

.... (2)

b) What is ultrasound?

....

.... (2)

c) Describe how ultrasound is used to produce an image of a body organ.

....

....

....

.... (3)

d) Suggest **two** advantages of using ultrasound rather than an X-ray photograph to examine a fetus.

....

.... (2)

TOTAL 9

3 The diagram represents a wave travelling along a rope.

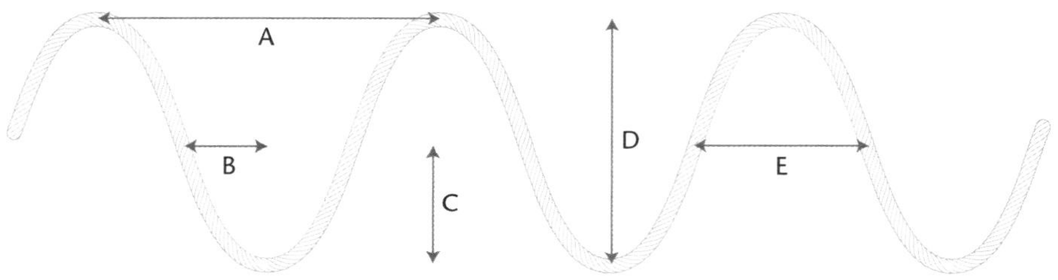

A
B
C
D
E

a) i) Which of the distances marked on the diagram represents the amplitude of the wave?

.... (1)

ii) Which of the distances marked on the diagram represents the wavelength of the wave?

.... (1)

iii) When a wave of frequency 3 Hz is sent along the rope, the wavelength is 0.6 m. Calculate the speed of the wave along the rope.

....

....

.... (3)

b) Some waves are transverse and some are longitudinal. Ropes can only transmit transverse waves.

i) Describe the difference between a transverse wave and a longitudinal wave.

.... (2)

ii) Give one other example of a transverse wave.

.... (1)

TOTAL 8

ANSWERS ON PAGE 95 ANSWERS ON PAGE 95 ANSWERS ON PAGE 95 ANSWERS ON PAGE 95

4 The electromagnetic spectrum is a family of waves with wavelengths ranging from approximately 1×10^{-15} m to hundreds of metres. The waves with shorter wavelengths than visible light are the most hazardous to humans.

a) Name two types of electromagnetic radiation with a wavelength shorter than light.

....

.... (2)

b) For each of the two types of radiation that you named in a) describe briefly:

i) the origin of the radiation

ii) one use of the radiation

iii) one possible danger of the radiation

iv) one precaution that should be taken to protect people who come into contact with that radiation.

....

....

....

....

....

....

....

....

....

.... (8)

TOTAL 10

---

5 The diagram shows light passing along an optical fibre.

a) Name the effect that takes place when light strikes the boundary of the fibre.

.... (1)

b) What is the condition for this to take place?

....

.... (2)

c) Information can be sent along optical fibres using either analogue or digital signals.

i) Describe the difference between an analogue and a digital signal.

....

....

.... (2)

ii) Give **two** advantages of using digital signals instead of analogue signals.

....

.... (2)

ANSWERS ON PAGE 95 ANSWERS ON PAGE 95 ANSWERS ON PAGE 95 ANSWERS ON PAGE 95

d) An optical fibre communication system uses radiation from a laser.
The wavelength of the radiation is 8.4×10^{-7} m.
Which part of the electromagnetic spectrum does this radiation fit into?

.... ①

TOTAL 8

---

6 After an earthquake, two types of wave travel through the body of the Earth.
These are P waves (longitudinal waves) and S waves (transverse waves).
The diagram shows how these waves pass through the Earth.

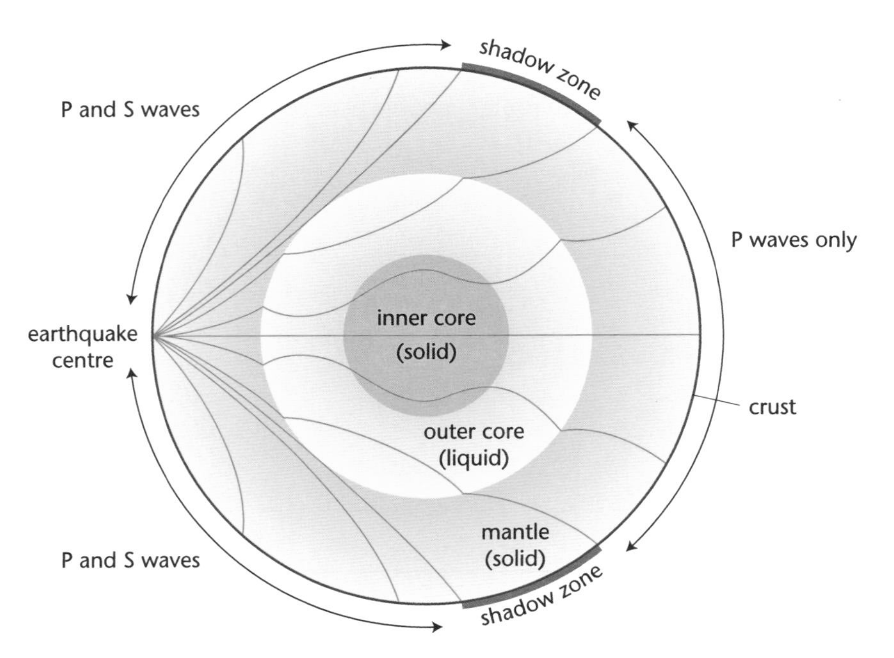

a) What causes the waves to change direction as they pass through the mantle?

.... ①

b) Which waves are detected in the shadow zone?

.... ①

c) Explain how the detection of these waves gives evidence that part of the Earth's core is liquid.

....

....

....

.... ④

TOTAL 6

ANSWERS ON PAGE 95 ANSWERS ON PAGE 95 ANSWERS ON PAGE 95 ANSWERS ON PAGE 95

QUESTION BANK ANSWERS

1 a)i) Diffraction 1
ii) The size of the opening 1
The wavelength 1

Examiner's tip

Frequency is not an alternative answer to wavelength. Electromagnetic waves and sound waves of similar wavelengths are diffracted by similar amounts, but they have very different frequencies.

iii) The width of a doorway is approximately one wavelength for sound 1
But it is many wavelengths for light 1

Examiner's tip

Always stress the width of the opening compared to the wavelength of the waves when answering questions about diffraction.

b)i) Wavelength = speed ÷ frequency 1
= 330 m/s ÷ 15 000 Hz 1
= 0.022 m. 1
ii) The wavelength is close to the size of the small speaker or the large speaker is many times the wavelength 1
The smaller speaker gives more spreading 1

2 a) X-rays are produced by X-ray machines 1
Gamma rays are emitted from an atomic nucleus 1

Examiner's tip

X-rays and gamma rays are both high frequency, short wavelength electromagnetic radiation. The only difference between them is where they come from.

b) High frequency or short wavelength 1
Compression or longitudinal wave 1

Examiner's tip

At this level a greater depth of answer than 'a high frequency sound wave' is required for 2 marks.

c) Sound from a vibrating crystal passes into the body
It is reflected at tissue boundaries
It is detected by the crystal
The information is processed by a computer
any three 3
d) Ultrasound does not harm the fetus 1
It can be used to see the fetus moving 1

Examiner's tip

X-rays can cause cell and tissue damage and could be very harmful to a fetus.

3 a)i) C 1

Examiner's tip

The common error is to describe D as the amplitude. D is twice the amplitude, which is the greatest displacement from the mean position.

ii) A 1

Examiner's tip

Any distance that is precisely one complete cycle is the wavelength, no matter what the starting point.

iii) speed = frequency × wavelength 1
= 3 Hz × 0.6 m 1
= 1.8 m/s 1
b)i) Transverse – the vibrations are at right angles to the direction of travel 1
Longitudinal – the vibrations are parallel to the direction of travel 1

Examiner's tip

The key words here are 'at right angles' and 'parallel'.

ii) X-rays, gamma rays, ultra violet, light, infra-red, microwaves, radio any one 1

4 a) Ultra-violet, X-rays and gamma rays
any two, 1 mark each 2
b) **ultra-violet**
i) Emitted by mercury vapour lamps or comes from the Sun 1
ii) Used on sun beds or to reveal security marking or in fluorescent lights 1
iii) Can cause skin cancer 1
iv) Keep the skin covered 1

X-rays
i) Come from X-ray tubes 1
ii) Used to examine bones or other body organs or to treat cancer 1
iii) Cause damage to healthy cells 1
iv) Avoid the X-ray beam or wear lead-lined protective clothing 1

gamma rays
i) Come from unstable nuclei 1

ii) Used for sterilising medical instruments or examining welds or treating cancer ❶
iii) Cause damage to healthy cells ❶
iv) Do not get close to the source, or wear lead-lined protective clothing, or monitor exposure to the radiation to make sure that 'safe' levels are not exceeded ❶

Examiner's tip

Extended writing is a key feature of science examinations. You will be expected to be able to write a description or put forward a logical argument where several points have to be made. Candidates often fail to gain marks on this type of question because they write at length about one point rather than the several that are asked for. Try to make your answers clear and concise, as space on the examination paper is often limited.

❺ a) Total internal reflection ❶

Examiner's tip

No marks are awarded for an answer that just states 'reflection'. An optical fibre can carry information over a large distance because all the light is reflected if it meets the boundary – none escapes.

b) The angle of incidence is greater than ❶
The critical angle ❶
c) i) An analogue signal can have any value ❶
But a digital signal can only be either 0 or 1 ❶
ii) Noise is easily removed from a digital signal ❶
A digital signal can carry more information ❶

Examiner's tip

CDs and computer disk drives store information as a series of 0s and 1s.

d) Infra-red ❶

Examiner's tip

You should know typical wavelengths and frequencies of the waves in the electromagnetic spectrum. If you remember the wavelengths, you can work out the frequencies using $v = f \times \lambda$.

❻ a) They change in speed ❶

Examiner's tip

The change in speed, or refraction, is also responsible for the change in direction when light passes from air into water or glass at any angle other than along the normal line.

b) None ❶

Examiner's tip

You are not expected to know this, but you are expected to be able to interpret the diagram.

c) Only P-waves are detected immediately opposite the earthquake centre ❶
These must have travelled through a liquid ❶
Since longitudinal waves can travel through a liquid ❶
But transverse waves cannot ❶

FOR MORE INFORMATION ON THIS TOPIC ... SEE REVISE GCSE SCIENCE ... CHAPTER 11

CHAPTER 12

The Earth and beyond

To revise this topic more thoroughly, see Chapter 12 in *Letts Revise GCSE Science Study Guide.*

Try this sample question and then compare your answers with the Grade C and Grade A model answers on the next page.

a Stars and some planets appear as bright objects in the night sky.

Describe the difference in the way in which we see stars and the way in which we see planets. **[2]**

b Explain why the planet Pluto cannot be seen with the unaided eye but Venus can be seen clearly. **[2]**

c The diagram shows the stages in the evolution of a small star.

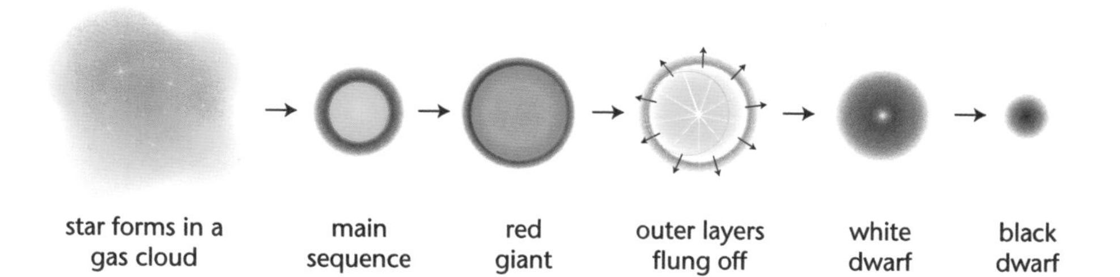

(i) Our Sun is in its main sequence.

Describe what is likely to happen to the Solar System when the Sun reaches the end of this stage. **[3]**

(ii) State the main difference between a white dwarf and a black dwarf. **[1]**

(Total 8 marks)

These two answers are at Grade C and A. Compare which one your answer is closest to and think how you could have improved it.

GRADE C ANSWER

Dave

a Stars are hot balls of gas on fire but planets are just lumps of rock. Stars give out light because they are very hot. ✓

Just one mark here for realising that stars give out light. Dave has the common misconception that stars are balls of gas on fire – they are not.

b Pluto is a long way from the Earth but Venus is nearer, so the light from it is brighter. ✗

No marks. Dave has not explained why the light from Venus appears brighter than that from Pluto.

c (i) The Sun will get bigger as it becomes a red giant. The planets will cool because they receive less energy from the Sun. ✓

Dave is awarded one mark because he has interpreted the diagram correctly in terms of the expansion of the Sun. He has not appreciated that it will absorb the Solar System.

(ii) They are different colours. ✗

No marks for this response. The question is about why they have different appearances.

2 marks = Grade C answer

Grade booster ⇢ move a C to a B

A grade B candidate should show a better understanding about how light becomes dimmer as it spreads out from a light source. Learning about the stages in the life cycle of a star would have enabled Dave to obtain more marks on part c.

GRADE A ANSWER

Sharma

a We see the light that stars give out but we can only see planets by the light that they reflect from the Sun and stars. ✓✓

This answer is correct and gains two marks.

b Pluto is a long way from the Sun so it doesn't get much light anyway, and then by the time that the light has travelled back to Earth it has spread out, making it very dim. ✓
Venus is closer to the Sun and the Earth, so it is lit up much brighter and the light doesn't spread out so much in travelling to Earth. ✓

Correct again – and explained very well. Two marks.

c (i) It will expand and get colder.
It will also absorb all the planets due to its enormous size. ✗

Sharma should have gained all three marks here, but her answer is imprecise. The question is about the Solar System and in her answer she uses the word 'it'. This is taken to refer to the Solar System. Had she begun her answer with 'The Sun', instead of 'It', she would have gained full marks.

(ii) The white dwarf is white because it is still emitting light. The black dwarf is too cold to give out any light, so it cannot be seen. ✓

Sharma gains the mark on this part of the question for a correct answer.

5 marks = Grade A answer

Grade booster ⇢ move A to A*

An A* candidate should have a good understanding of how stars are formed due to the effect of gravitational forces. At A* level a candidate should also be able to give precise, unambiguous answers.

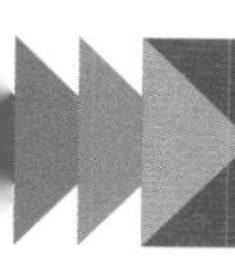

QUESTION BANK

1

a) i) Which word best describes the Earth's moon? Choose from the list.

asteroid **planet** **satellite** **star**

.... (1)

ii) It takes 365 days for the Earth to complete one orbit of the Sun.
It takes 29 days for the Moon to complete one orbit of the Earth.
How long does it take for the Moon to travel once around the Sun?

.... (1)

iii) Choose **two** words from the list that describe the force between the Earth and the Moon.

attractive **electrostatic** **gravitational** **magnetic** **repulsive**

.... (1)

b) The diagram shows the force that the Earth exerts on the Moon.

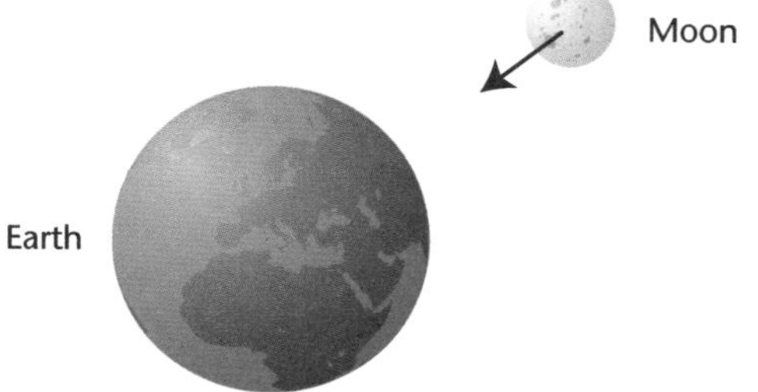

i) Add an arrow to the diagram that shows the Moon's pull on the Earth. (1)

ii) Which statement is correct?
A The size of the Earth's pull on the Moon is greater than that of the Moon's pull on the Earth.
B The size of the Moon's pull on the Earth is greater than that of the Earth's pull on the Moon.
C The size of the Moon's pull on the Earth is equal to that of the Earth's pull on the Moon.

.... (1)

c) Moon rocks that have been brought back to Earth show no evidence of life ever having existed on the Moon.
Suggest why it is probable that there has never been life on the Moon.

....

....

....

.... (3)

TOTAL 8

2 The table contains data about some of Jupiter's moons.

Moon	Distance from Jupiter in thousands of km	Orbital period in days	Radius in km	Density in g/cm³
Ganymede	1070	7.2	2631	1.9
Callisto	1880	16.7	2400	1.8
Io	422	1.8	1815	3.6
Europa	671	3.6	1569	3.0

a) Which is the largest of these four moons?

.... (1)

b) What type of force keeps the moons in orbit around Jupiter?

.... (1)

c) What is the relationship between the orbital period of a moon and its distance from Jupiter?

....

.... (2)

d) Io has a greater mass than Europa. Give **two** reasons why.

....

.... (2)

e) Io's volcanic eruptions reach heights of hundreds of kilometres.
Suggest **two** reasons why the eruptions from Io's volcanoes reach much greater heights than those from volcanoes on Earth.

....

.... (2)

TOTAL 8

---

3 The Universe is thought to have started with an enormous explosion. The first substances that formed after the explosion were hydrogen and helium. Heavier elements were later formed in stars.

a) Suggest how the heavier elements could have formed.

....

.... (2)

b) Explain why our Solar System could not have been formed at the same time as the first stars in the Universe. One mark is awarded for correct spelling, punctuation and grammar.

....

....

....

.... (3)

c) One theory about the origin of the Solar System is that it was formed from a spinning cloud of dust that was part of the remains of an exploding supernova.
Explain how this theory explains the fact that the innermost planets are denser than the outer planets.

....

.... (2)

d) The Universe is still expanding, but the rate of expansion is decreasing. Suggest why the rate of expansion is decreasing.

....

.... (2)

e) Explain how the expansion of the Universe supports the theory that it started as a single explosion.

....

.... (2)

TOTAL 11

ANSWERS ON PAGE 102 ANSWERS ON PAGE 102 ANSWERS ON PAGE 102 ANSWERS ON PAGE 102

4 The diagram represents the motion of the two innermost planets, Mercury (M) and Venus (V).

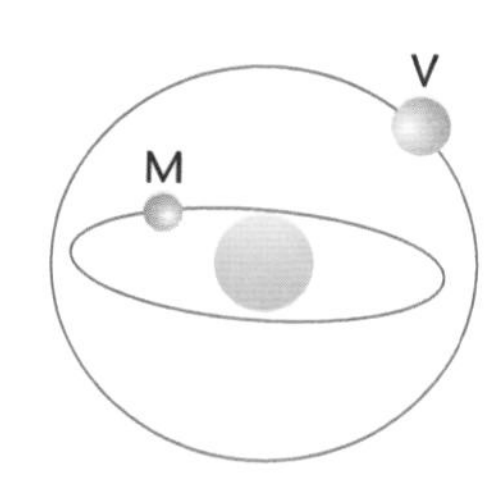

The table gives some data about these planets.

	Mercury	Venus
Mass in kg	3.3×10^{23}	4.9×10^{24}
Radius in m	2.4×10^{6}	6.1×10^{6}
Density in g/cm^3	5·4	5·3
Orbital period in Earth days	88	224
Distance from Sun compared to Earth	0.39	0.72
Mean orbital speed in km/s	48	35
Atmosphere	none	carbon dioxide with sulphuric acid clouds

Use the diagram and the information in the table to answer the questions that follow.

a) The speed of Venus in its orbit around the Sun hardly changes, but that of Mercury varies considerably. Suggest why this is.

....

.... (2)

b) Describe and explain how you would expect the speed of Mercury to change as it orbits the Sun.

....

....

.... (4)

c) Suggest **two** reasons why Venus takes longer than Mercury to complete one orbit of the Sun.

....

.... (2)

d) Venus is similar in size and mass to the Earth. What does this information tell you about the densities of the three innermost planets?

.... (1)

e) Mercury and the Earth's moon both show evidence of many collisions with meteors. These are thought to have occurred in the early part of the Solar System's lifetime. Suggest **three** reasons why there is little evidence of collisions on Earth and Venus.

....

....

.... (3)

f) Space probes that have landed on Venus to photograph the surface have had a very short operational life. Suggest **two** reasons for this.

....

.... (2)

TOTAL 14

QUESTION BANK ANSWERS

1 a) i) Satellite (1)

Examiner's tip

Any object that orbits another object is a satellite.

ii) 365 days. (1)

Examiner's tip

The Moon orbits the Earth but it also orbits the Sun with the same time period as that of the Earth.

iii) Attractive and gravitational
both required (1)

b)i) The arrow should point from the centre of the Earth towards the centre of the Moon (1)

ii) C

Examiner's tip

Forces act between objects. The forces on each of the two objects are always equal in size and opposite in direction.

c) The Moon has no atmosphere (1)
There is no known water on the Moon (1)
These are essential conditions for life to exist (1)

2 a) Ganymede (1)

b) Gravitational (1)

c) As the distance from Jupiter increases (1)
The orbital period increases (1)

Examiner's tip

Here is another example of where you need to take care to give a precise answer. An answer such as 'the orbital period increases with distance' would only gain one mark. For both marks, it needs to be clear that the period increases as the distance increases.

d) Io is denser than Europa (1)
Io is larger than Europa (1)

e) Io's gravitational field strength is less than the Earth's (1)
There is no atmosphere (or water) to exert a resistive force (1)

Examiner's tip

Much of this question is concerned with interpreting data given in the table. This type of question is common at GCSE on this topic.

3 a) Hydrogen and helium nuclei were forced together at high speeds (1)
They fused to form the nuclei of heavier elements (1)

Examiner's tip

Atoms of different elements are made in stars by nuclear fusion.

b) The Earth is rich in elements that are heavier than hydrogen and helium (1)
These did not exist when the first stars formed (1)
Communication mark – answer written in sentences with correct punctuation and grammar (1)

Examiner's tip

The existence of atoms of 'heavy' elements provides evidence that the Solar System is much younger than the Universe.

c) The lighter materials would have moved to the outside of the spinning cloud (1)
Leaving the heavier materials on the inside (1)

d) All the objects in the Universe attract each other (1)
This gravitational attractive force is slowing down the rate of expansion (1)

Examiner's tip

The rate of expansion is slowing down because the attractive forces between the galaxies are in the opposite direction to their motion.

e) The path of the galaxies can be traced back (1)
They appear to have started from a common point in space (1)

Examiner's tip

This is an important piece of evidence that supports the 'Big Bang' theory of the origin of the Universe.

FOR MORE INFORMATION ON THIS TOPIC ... SEE REVISE GCSE SCIENCE ... CHAPTER 12

4 a) Venus has a circular orbit, so it is a constant distance from the Sun (1)
Mercury's orbit is elliptical, so its distance from the Sun varies (1)

Examiner's tip

The speed of a planet in its orbit depends on the strength of the Sun's gravitational field only. This in turn depends on the distance between the planet and the Sun. A common misconception at GCSE is that the more massive planets travel more slowly around the Sun compared to the less massive planets.

b) As Mercury gets nearer to the Sun it speeds up (1)
Because the gravitational pull of the Sun increases (1)
As it moves further away from the Sun it slows down (1)
Because the gravitational pull of the Sun decreases (1)

c) Venus travels slower in its orbit than Mercury does (1)
It also has further to travel to complete an orbit (1)

Examiner's tip

The orbit time of planets increases with increasing distance from the Sun. The further a planet is from the Sun the further it has to travel to complete an orbit and the slower its orbital speed.

d) They all have similar densities

Examiner's tip

You can tell from the table that Mercury and Venus have similar densities. The information that you are given in this question also tells you that the Earth has a similar density to that of Venus. This reinforces the point that you need to read questions carefully so that you do not miss important information.

e) The Earth and Venus both have atmospheres (1)
They also both have clouds (1)
Wind and rain would have eroded craters (1)

f) The atmosphere is acidic and very corrosive (1)
It is also likely to be very hot (1)

Examiner's tip

Venus is closer to the Sun than the Earth, so it receives more radiation on each square metre. The carbon dioxide atmosphere helps to keep the planet hot, as a result of the greenhouse effect.

CHAPTER 13

Energy resources and energy transfer

To revise this topic more thoroughly, see Chapter 13 in *Letts Revise GCSE Science Study Guide*.

Try this sample GCSE question and then compare your answers with the Grade C and Grade A model answers on the next page.

Most of the UK's electricity is generated by burning the fossil fuels, coal and gas. Some is generated from nuclear fuel and a small amount comes from renewable resources such as wind and hydroelectric power.

a The diagram compares the energy flow through a coal-burning and a gas-burning power station.

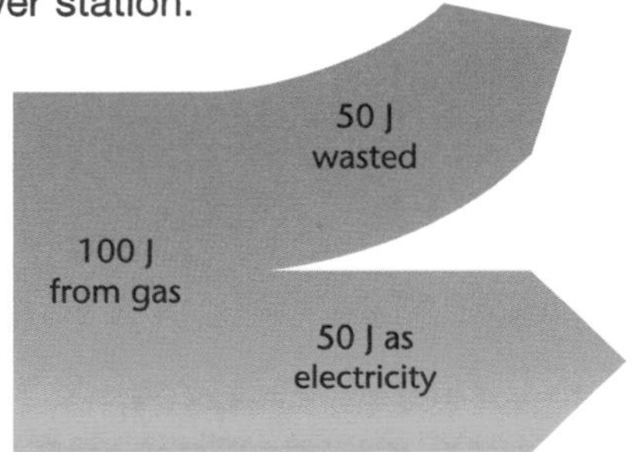

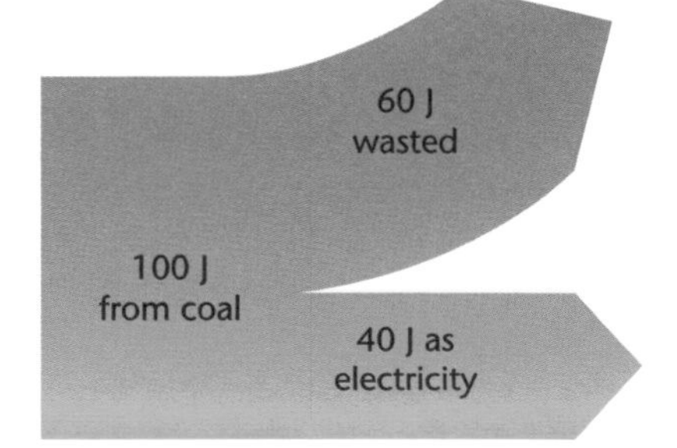

(i) Which power station is more efficient?
Explain how you can tell from the diagram. **[2]**

(ii) In recent years some older, coal-burning power stations have closed and been replaced by gas-burning power stations.
Explain how this has affected the environment. **[3]**

b Electricity is generated in a power station by an electromagnet rotating within copper coils.

(i) What causes a voltage to be induced in the copper coils? **[1]**

(ii) The speed of rotation of the electromagnet is often greater at night than during the day.
How does this affect the output voltage from the generator? **[2]**

(iii) Electricity is generated at 25 000 V.
For distribution to consumers, the voltage is increased to 400 000 V.
Explain why. **[3]**

(Total 11 marks)

These two answers are at Grade C and A. Compare which one your answer is closest to and think how you could have improved it.

GRADE C ANSWER

Robina

a (i) The gas one because less energy is wasted. ✓✓

(ii) It's colder because there's less waste energy to warm it up. ✓✓

b (i) The magnetic field of the electromagnet ✗

(ii) The voltage is bigger ✓

c So that the wires don't get as hot. ✓

This way, more electricity gets to houses. ✗

This answer scores full marks – two out of two. Robina has identified the power station correctly and given the correct reason.

Two marks out of three here. Robina has given one correct effect and the reason for it. She has not made any statement about the comparative amounts of carbon dioxide that the power stations emit.

Robina's answer has missed the important point: it is the changing magnetic field that causes the induced voltage. No marks are awarded for this answer.

One mark out of two. The effect on the voltage has been given correctly, but she has not appreciated that the frequency is also affected.

One mark for stating that less energy is lost as heat.

6 marks = Grade C answer

Grade booster ···> move a C to a B

A grade B candidate should get the marks in c by putting forward a logical argument. This argument should state that higher voltages mean that smaller currents are used, which leads to less energy wasted as heat.

GRADE A ANSWER

Mike

a (i) There is less waste in the gas-fired power station, so this is more efficient. ✓✓

(ii) Gas-fired power stations disperse less heat to the environment. ✓ They also produce less carbon dioxide, which is a greenhouse gas. ✓ Carbon dioxide gives people skin cancer, so the less there is, the better.

b (i) The magnetic field through the coils is changing all the time because the electromagnet is turning round. ✓

(ii) It gets bigger. ✓

c With a higher voltage, the current is smaller, so thinner wires are needed. ✓✓

Two out of two for a very good answer.

This answer starts off well. It gains 1 mark for stating that less heat is passed to the environment and a second mark for appreciating that gas-fired power stations produce less carbon dioxide than coal-fired stations do. The last part of the answer shows confusion between the effects of greenhouse gases and those of ozone layer depletion, so gains no marks.

A good answer – one mark awarded.

Like Robina, Mike gains a mark for the bigger voltage but has missed out on the mark for stating that the frequency also increases.

Two marks out of three. Mike has not stated here that the lower the current, the less the energy losses.

8 marks = Grade A answer

Grade booster ···> move A to A*

A* candidates should be able to give a full explanation of why electricity is transmitted at a high voltage in c. They would also be expected to understand that increasing the speed of the electromagnet affects both the size of the voltage and its frequency in b (ii).

QUESTION BANK

1 Although most of the electricity generated in the UK comes from burning fossil fuels such as coal, oil and gas, a significant amount is generated using moving water; this is called 'hydroelectric power'.

a) Write down **three** advantages of using hydroelectric power instead of burning fossil fuels.

....

....

.... ③

b) What is the energy source for hydroelectric power?

.... ①

c) In a pumped storage system, water is pumped from a low reservoir to a high one at night when there is little demand for electricity. The water can be released, generating electricity as it goes back to the low reservoir, in order to satisfy peak demand. The diagram shows the arrangement of the reservoirs.

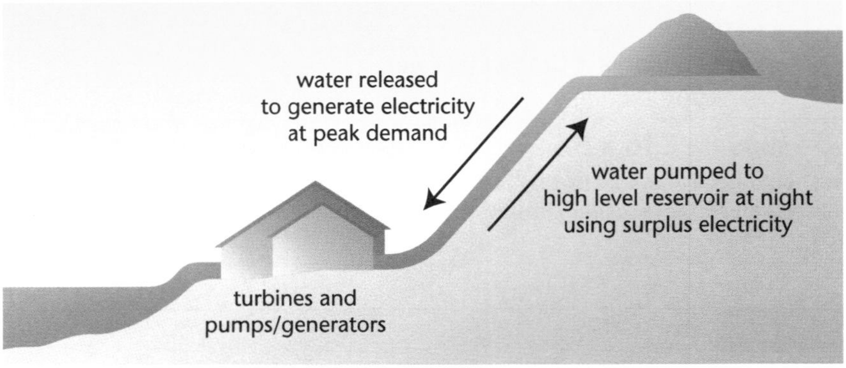

i) Describe the energy transfer when the power station is generating electricity. One mark is awarded for a clearly ordered answer.

....

....

.... ④

ii) In a pumped storage station, the maximum water flow downhill is 2.5×10^5 kg each second. This water falls through a height of 190 m. Calculate the energy lost by 2.5×10^5 kg of water in falling through a height of 190 m.
The strength of the Earth's gravitational field, $g = 10$ N/kg.

....

....

.... ③

iii) The energy from the water is transferred to electricity with an efficiency of 60%. Calculate the maximum power output of the power station.

....

....

.... ③

iv) Suggest **one** factor that limits the efficiency of a pumped storage station when it is generating electricity.

....

.... ①

ANSWERS ON PAGE 109 ANSWERS ON PAGE 109 ANSWERS ON PAGE 109 ANSWERS ON PAGE 109

v) Pumped storage stations do not produce electricity. It takes more energy to pump the water uphill than is recovered when it moves down.
Suggest **two** advantages of using pumped storage rather than building more power stations to satisfy peak demand.

....

.... ②

TOTAL 17

---

2 As an alternative to lifting a heavy load onto the back of a lorry, a ramp can be used to pull it up a slope.

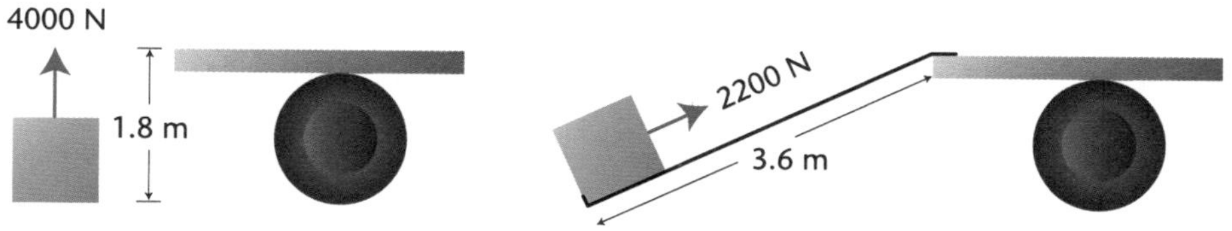

4000 N
1.8 m
2200 N
3.6 m

a) Calculate the work done when the load is lifted onto the lorry.

....

....

.... ③

b) Calculate the work done by dragging the load up the ramp.

....

.... ②

c) The answer to a) is the gain in gravitational potential energy of the load in each case. Explain why more work is needed to drag the load up the ramp than to lift it, and suggest where the extra energy goes to.

....

....

.... ③

d) State one advantage of using the ramp instead of lifting the load.

.... ①

TOTAL 9

---

3 Some takeaway shops put hot food in aluminium foil containers. Others use expanded polystyrene.

aluminium foil

polystyrene

a) Why is it important that both types of container are fitted with lids?

....

.... ②

b) By what method does energy from the food pass through the container walls?

.... ①

c) Which type of container is better for keeping food hot?
Explain your answer.

....

.... ②

d) Some takeaway foods are wrapped in several sheets of paper.
Explain how this provides effective insulation.

....

.... ②

TOTAL 7

---

4 Two coils of wire are wound on an iron core.
One coil is connected to a battery and a switch.
The other coil is connected to a sensitive ammeter that can detect current in either direction.

sensitive ammeter

a) What happens to the ammeter pointer when:

i) the switch is closed

.... ①

ii) the switch remains closed

.... ①

iii) the switch is opened?

....

.... ②

b) The left-hand coil is now connected to an a.c. supply and the right-hand coil is connected to a lamp.

Explain why the lamp lights.

....

....

.... ③

c) In an arrangement similar to that in b), a 3 V a.c. supply is connected to a 60-turn coil of wire.
The second coil is connected to a 12 V lamp, which lights at its normal brightness.
How many turns of wire are on the second coil?

....

....

.... ③

TOTAL 10

ANSWERS ON PAGE 109 ANSWERS ON PAGE 109 ANSWERS ON PAGE 109 ANSWERS ON PAGE 109

QUESTION BANK ANSWERS

1 a) Hydroelectric power is renewable; it does not use up any of the Earth's energy resources **1**
It does not produce carbon dioxide or other greenhouse gases **1**
It does not leave ash or produce sulphur dioxide **1**

Examiner's tip

Many candidates confuse the effects of pollution caused by emission of gases into the atmosphere. Carbon dioxide, water vapour and methane all contribute to the greenhouse effect and sulphur dioxide contributes towards acid rain.

b) The Sun **1**

Examiner's tip

The Sun provides the majority of the Earth's energy sources. In the case of hydroelectricity, it provides the energy to evaporate the water that falls as rain. The exception is the tides, which are mainly due to the Moon's rotation around the Earth.

c) i) Gravitational potential energy in the water **1**
Changes to kinetic energy as the water falls **1**
This is transferred to electricity in the generators **1**
+1 for a clearly ordered answer **1**

ii) change in gpe = mass × gravitational field strength × change in height **1**
$= 2.5 \times 10^5$ kg × 10 N/kg × 190 m **1**
$= 4.75 \times 10^8$ J **1**

Examiner's tip

When calculating the change in gravitational potential energy, remember to use the vertical change in height rather than the distance that an object moves up or down a slope.

iii) max power output = efficiency × power input **1**
$= 0.6 \times 4.75 \times 10^8$ J/s **1**
$= 2.85 \times 10^8$ W **1**

iv) Not all the kinetic energy can be removed from the water; it has to have some movement when it leaves the generators **1**

v) Pumped storage stations can be brought into operation very rapidly
They use electricity generated at night when demand is low and it is readily available
They do not cause any atmospheric pollution
any two, 1 mark each **2**

2 a) work = force × distance moved in direction of force **1**
= 4000 N × 1.8 m **1**
= 7200 J **1**

b) work = 2200 N × 3.6 m **1**
= 7920 J **1**

c) More work has to be done when dragging because of friction forces **1**
The extra energy is transferred to heat **1**
The heat is in the ramp and the load **1**

Examiner's tip

Whenever two surfaces rub against each other some work has to be done against the friction forces that oppose the motion; this causes heating of the surfaces that are rubbing.

d) The ramp enables a smaller force to be used **1**

3 a) Hot food loses energy through evaporation and convection **1**
The lids prevent the vapour and warm air from escaping **1**

Examiner's tip

Evaporation and convection are the main methods of energy loss from hot food and drink. This is why covering hot food with aluminium foil keeps it hot, even though the aluminium is a good thermal conductor.

b) Conduction **1**

c) Polystyrene **1**
It is an insulator, so less energy is lost by conduction **1**

d) Air is trapped between the layers of paper **1**
Trapped air is a good thermal insulator **1**

Examiner's tip

Many methods of insulation, for example cavity wall insulation, loft insulation and double glazing, rely on using air as the insulator. If the air is trapped it cannot remove energy through convection currents.

Energy

4 a) i) It moves/shows a reading and then returns to zero (1)
ii) The pointer stays at zero (1)
iii) It moves/shows a reading and then returns to zero (1)
The movement/reading is in the opposite direction to that in a) i) (1)

Examiner's tip

A voltage or current is only induced in the right-hand coil when the magnetic field is changing. This happens when a direct current is switched on or off. Switching the current off causes an induced voltage in the opposite direction to when it is switched on.

b) The current is continually changing (1)
So the magnetic field is continually changing (1)
A voltage is induced whenever the magnetic field changes (1)

Examiner's tip

Always emphasise the changing magnetic field when explaining the effects of electromagnetic induction.

c) $\frac{\text{primary voltage}}{\text{secondary voltage}} = \frac{\text{number of primary turns}}{\text{number of secondary turns}}$ (1)

$\frac{3\,V}{12\,V} = \frac{60}{N_s}$ (1)

$N_s = 240$ (1)

Examiner's tip

This type of question is easier to answer by using ratios than by using the transformer equation. If the secondary voltage is four times that of the primary, the secondary coil must have four times as many turns as the primary coil.

FOR MORE INFORMATION ON THIS TOPIC ... SEE REVISE GCSE SCIENCE ... CHAPTER 13

CHAPTER 14

Radioactivity

To revise this topic more thoroughly, see Chapter 14 in *Letts Revise GCSE Science Study Guide.*

Try this sample GCSE question and then compare your answers with the Grade C and Grade A model answers on the next page.

Technetium-99 is a radioactive isotope used in medicine. It decays by emitting gamma radiation.

a What is gamma radiation? **[2]**

b The activity of a sample of technetium-99 is measured at intervals of two hours, starting when the sample has been prepared. The graph shows the results.

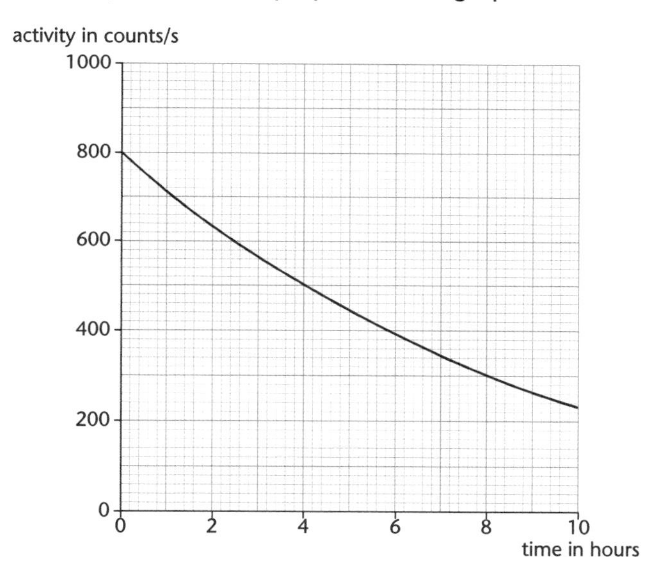

(i) The sample was injected into a patient one hour after being prepared.
What was the activity of the sample when it was injected into the patient? **[1]**

(ii) Use the graph to determine the half-life of technetium-99.
Show how you obtain your answer. **[3]**

(iii) Estimate the activity of the sample one day after it is injected into the patient. **[3]**

c Technetium-99 can be used to monitor the blood flow in organs such as the liver. After the patient has been injected, a camera is used to photograph the patient from the outside. Explain why a gamma emitter is used rather than a substance that emits alpha or beta particles. **[3]**

(Total 12 marks)

These two answers are at Grade C and A. Compare which one your answer is closest to and think how you could have improved it.

GRADE C ANSWER

Basil

a Gamma radiation is an electromagnetic wave ✓

b (i) 710 counts/s ✓

(ii) It takes nearly 6 hours for the counts to halve from 800 to 400. ✓

(iii) None, because two half-lives is 12 hours so after this there's none left. ✗

c Gamma isn't dangerous radiation and it shows up better on photographs. Alpha and beta are more dangerous to the patient because they would just stay around inside the body. ✓

One mark out of two here. Basil has only made one point in his answer and there are two marks allocated.

One mark for reading the graph correctly.

Basil shows in his answer that he knows how to read the half-life from the graph, so he gets one mark. The precision of his answer is not good enough for the second mark and he has not realised that he needs to take more than one reading and work out the average.

No marks here. Basil has the common misconception that if half the undecayed nuclei decay in one half-life, the other half decay in the same time.

Only one mark here for appreciating that alpha and beta radiations would not penetrate the body. He needs to be more specific about the effects and dangers of alpha and beta radiation. His statement that 'gamma isn't dangerous' is also wrong.

4 marks = Grade C answer

Grade booster ····› move a C to a B

A grade B candidate should be able to take precise readings from a graph and be able to understand the meaning of half-life.

GRADE A ANSWER

Tamsin

a Gamma radiation is electromagnetic radiation that has a very short wavelength. ✓✓

b (i) 710 counts/s ✓

(ii) The activity at the start is 800 counts/s. I've drawn a line across from 400 counts/s and it meets the curve at a time of 5.8 hours, so this is the half-life. ✓✓

(iii) One day is 24 hours and this is more than four half-lives. I think it will have all gone by this time. ✓

c Gamma radiation is the only one out of the three that can pass through the body to the camera. ✓ The others would be stopped inside the body. ✓

Tamsin is awarded two marks here for making two correct statements about gamma radiation.

Correct – one mark.

Correct again. Note how Tamsin has explained how she obtained her value for the half-life. Two marks. However, Tamsin has not taken an additional reading from the graph to enable her to work out an average value of the time taken for the rate of decay to halve.

Tamsin has worked out correctly that after one day the sample has been through approximately four half-lives. She is only awarded one mark because she has not been able to use this to work out the activity of the sample.

Two marks here – one for stating that gamma is the only one of the three types to pass through the body and one mark for stating that the others would be absorbed. Tamsin has not stated how this is potentially harmful to the patient.

8 marks = Grade A answer

Grade booster ····› move A to A*

An A* candidate should be able to use a graph to work out an average value for a half-life, be able to carry out half-life calculations and to explain why alpha and beta radiations are dangerous when used inside the body.

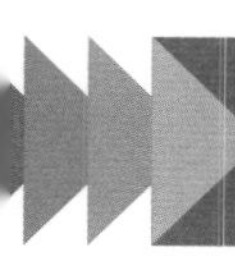

QUESTION BANK

1

a) A Geiger–Müller tube is connected to a counter. It measures the background radiation each 60 s for ten successive time intervals. The table shows the results.

Recording	1	2	3	4	5	6	7	8	9	10
Number of counts in 60 s	62	59	69	67	61	68	68	70	72	64

i) Name **two** sources of background radiation.

....

.... ②

ii) Explain why there is variation in the number of counts recorded in successive intervals.

.... ①

iii) Calculate a value for the average background radiation in counts/s.

.... ①

b) The same Geiger–Müller tube is used to measure the activity of a radioactive liquid containing protactinium-234, an isotope that has a half-life of around one minute. Readings of the activity are recorded at 10 s time intervals for a period of 2 minutes. The table shows the results.

Time in s	0	10	20	30	40	50	60	70	80	90	100	110	120
Count rate in counts per s	32.2	30.0	27.7	25.2	22.6	20.9	19.2	17.3	15.9	14.8	13.4	12.1	11.0
Corrected count rate in counts per s													

i) Use your answer to a) iii) to complete the data in the table. ①

ii) Use the data to plot a graph of corrected count rate against time. ④

iii) Use your graph to obtain a value for the half-life of protactinium-234. Explain how you arrive at your answer.

....

.... ②

TOTAL 11

ANSWERS ON PAGE 115 ANSWERS ON PAGE 115 ANSWERS ON PAGE 115 ANSWERS ON PAGE 115

2 $^{12}_{6}C$ and $^{14}_{6}C$ are both isotopes of carbon.

a) i) Write down **one** similarity about the nucleus of each isotope.

.... (1)

ii) Write down **one** difference in the nucleus of these isotopes.

.... (1)

b) $^{14}_{6}C$ is radioactive. It decays by emitting a beta particle.

i) Describe a beta particle.

.... (2)

ii) Which part of the atom emits the beta particle?

.... (1)

c) $^{14}_{6}C$ is present in all living materials and in all materials that have been alive. It decays with a half-life of 5730 years.

i) Explain the meaning of the term half-life.

....

.... (2)

ii) The activity of a sample of wood from a freshly-cut tree is measured to be 80 counts/s. Estimate the activity of the sample after two half-lives have elapsed.

.... (1)

iii) The age of old wood can be estimated by measuring its radioactivity. Explain why this method cannot be used to work out the age of a piece of furniture made in the nineteenth century.

....

.... (2)

TOTAL 10

---

3 Sodium-24 is a radioactive form of sodium that emits gamma radiation. It has a half-life of 15 hours.

a) Explain the meaning of the terms:

radioactive ....

.... (2)

gamma radiation ....

.... (2)

half-life ....

.... (2)

b) Sodium-24 can be used to trace leaks in underground water pipes. Radioactive salt is dissolved in the water before it enters the pipe.

i) Describe how a leak can be detected. One mark is awarded for spelling, punctuation and grammar.

....

....

.... (3)

ii) Explain how the properties of sodium-24 make it suitable for this purpose.

....

.... (2)

TOTAL 11

ANSWERS ON PAGE 115 ANSWERS ON PAGE 115 ANSWERS ON PAGE 115 ANSWERS ON PAGE 115

QUESTION BANK ANSWERS

1 a)i) The air, the ground, rocks, the Sun, waste from nuclear power stations, radiation from testing of nuclear weapons, radioactive materials used in medicine
any two 1 mark each **2**

ii) Radioactive decay is a random process **1**

Examiner's tip

The decay of an unstable nucleus cannot be predicted; it could occur at any time, which is why the process of radioactive decay is described as 'random'.

iii) 1.1 counts/s **1**

b)i) The completed table should show 1.1 counts/s deducted from the count rate to give the corrected count rate:

Corrected count rate in counts per s
31.1 28.9 26.6 24.1 21.5 19.8 18.1
16.2 14.8 13.7 12.3 11.0 9.9

ii) The completed graph should look like this

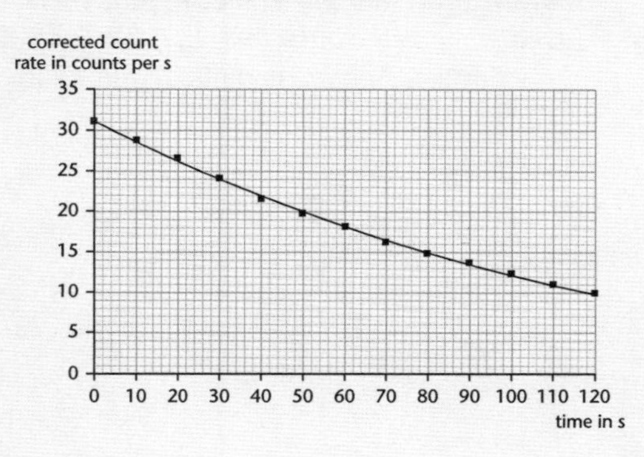

Award marks for:
Suitable scales and correct labels on the axes **1**
Correct plotting of all the points **2**
Drawing a smooth curve **1**

Examiner's tip

Because of the random nature of radioactive decay, all the points cannot be expected to lie precisely on a smooth curve. When drawing the curve, try to balance out the number of points above the curve with the number of points below the curve.

iii) Answer in the range 70 to 75 s **1**
Explanation to show that more than one pair of readings has been used;
e.g. 74 s to decay from 30 counts/s to 15 counts/s and 71 s to decay from 20 counts/s to 10 counts/s gives an average of 73 s **1**

Examiner's tip

Because of the random nature of radioactive decay, the half-life is defined as the average time for half the undecayed nuclei to decay. You need to take more than one reading from the graph to be able to work out an average.

2 a)i) They have the same number of protons **1**

ii) They have different numbers of neutrons **1**

Examiner's tip

Isotopes of the same element must have the same number of protons, so the atomic number (the lower number on the left of the symbol) is the same. They have different numbers of neutrons, giving the nuclei different mass numbers (the upper number on the left of the symbol).

b)i) Beta particle is a high-energy OR fast-moving **1**
Electron **1**

ii) The nucleus **1**

Examiner's tip

A common error is for candidates to state that the beta particle is emitted from the electrons around the nucleus, since the nucleus does not contain any electrons. The emission of a beta particle occurs when a neutron decays into a proton and an electron.

Radioactivity

c) i) Half-life is the average time ❶
For the number of undecayed nuclei to halve ❶
ii) 20 counts/s ❶

Examiner's tip

A common misunderstanding amongst GCSE candidates is that after two half-lives all the nuclei have decayed. This is not the case; on the average half should decay during one half-life and then half of these (i.e. one quarter of the original nuclei) during the next half-life and so on.

iii) One hundred years is a very short time compared to the half-life ❶
So the rate of decay would show no measurable change ❶

Examiner's tip

Radiocarbon dating is suitable for dating objects that are thousands of years old in order that a measurable difference between the activity of the carbon and that of carbon in new objects can be recorded.

❸ a) Radioactive – atoms have unstable nuclei ❶
They emit radiation ❶
Gamma radiation – short wavelength or high energy or high frequency ❶
Electromagnetic radiation ❶
Half-life – the average time ❶
Taken for the number of undecayed nuclei to halve ❶

Examiner's tip

A common misconception is that the half-life of a radioactive material is half its life, or half the time it takes to decay.

b) i) Use a Geiger–Müller tube or photographic film to detect the radiation ❶
The leak is where radiation level is highest or increases ❶
Quality of written communication – answer written in sentences with correct punctuation and grammar ❶
ii) It emits gamma radiation which can travel through soil ❶
The half-life is long enough for the leak to be detected ❶

Examiner's tip

When considering whether a radioactive material is suitable for a particular purpose, you need to consider both the type of radiation emitted and the half-life. The type of radiation to choose depends on whether it needs to be easily absorbed or not and the half-life to choose depends on whether it needs to become inactive quickly or stay radioactive for a long time. A short half-life is needed for a tracer used in medicine but a long half-life is needed for a thickness gauge in a paper mill.

FOR MORE INFORMATION ON THIS TOPIC ... SEE REVISE GCSE SCIENCE ... CHAPTER 14

Centre number
Candidate number
Surname and initials

Letts Examining Group

General Certificate of Secondary Education

Science: Biology Paper

Higher tier

Time: 1 Hour

Instructions to candidates

Answer all questions in the spaces provided.

The number of marks for each question is given in brackets at the end of each part.

These marks are a guide for the detail required in each answer.

Use a sharp pencil when drawing a graph or diagram.

In certain questions extra marks are available for the quality of your written answer.

Allocate a few minutes towards the end of the examination to check your answers.

Information for candidates

The number of marks available is given in brackets **[2]** at the end of each question or part question.

The marks allocated and the spaces provided for your answers are a good indication of the length of answer required.

For Examiner's use only	
1	
2	
3	
4	
5	
6	
7	
Total	

Leave blank

1 The diagram shows a section through a human eye.

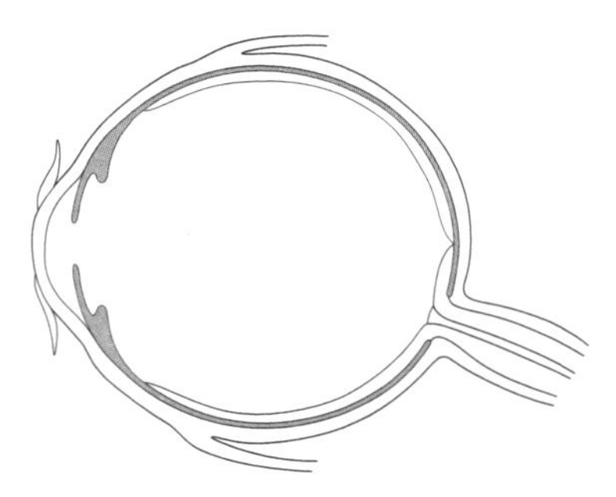

(a) **(i)** Label with **A** the part of the eye that carries nerve impulses to the brain.

(ii) Label with **B** the part of the eye that contains light sensitive cells.

(iii) Label with **C** the part of the eye that adjusts the size of the pupil. **[3]**

(b) The diagram shows light rays passing through the eye from a near object.

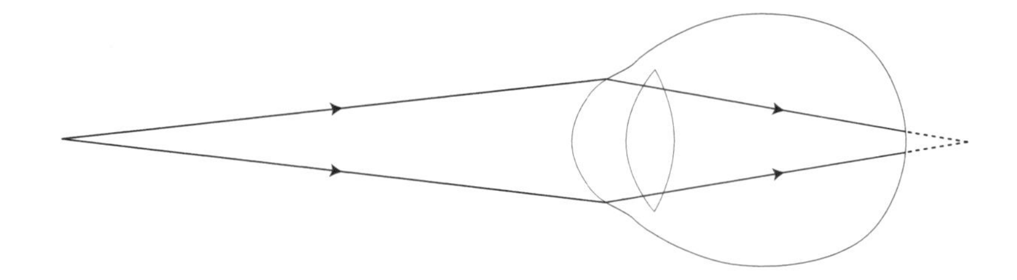

This person is long sighted.
Explain why they cannot see the object clearly.

....

....

.... **[2]**

(c) The light sensitive cells in the eye are called rods and cones.
Write down **two** differences between rods and cones.

1 ....

2 ....

.... **[2]**

(Total 7 marks)

Leave blank

2 Freshwater shrimps are small animals that live in streams.

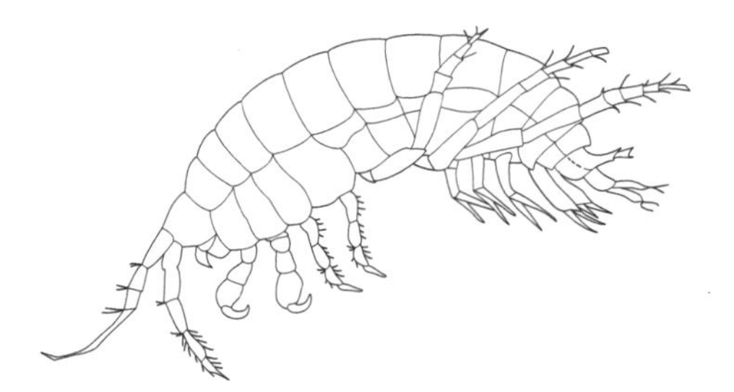

(a) The freshwater shrimp needs a regular supply of oxygen from the water.
The faster the water flows then the more oxygen it contains.
The shrimp uses this oxygen for respiration.

(i) Complete the following balanced equation of this respiration.

$6O_2$ + ⟶ + $6CO_2$ + energy **[2]**

(ii) Write down **one** way in which the shrimp uses the energy that is produced.

.... **[1]**

(b) Joshua decided to try to work out where the freshwater shrimps prefer to live.
He sampled 1 m^2 areas of the stream in different locations using a net.
He then counted how many shrimps he caught in the different locations.
He also worked out the speed of the water in each part of the stream.
He plotted his results as a graph.

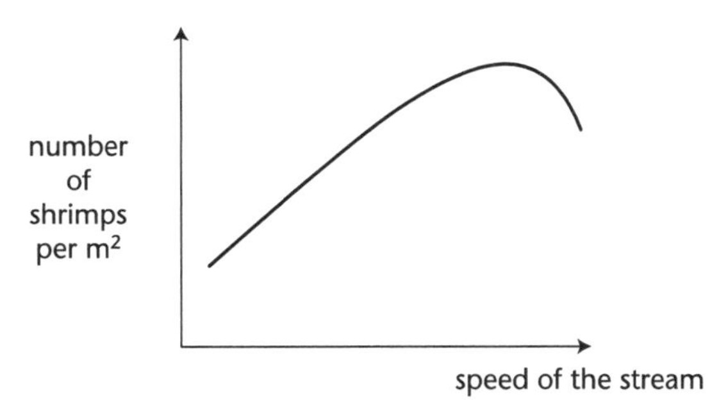

(i) Why did Joshua sample each area with his net for the same length of time?

....

.... **[2]**

(ii) Describe what Joshua's results tell him about where the shrimps like to live.

....

.... **[2]**

[turn over

Leave blank

(iii) Suggest why the shrimps like to live in these particular areas rather than others.

....

....

.... **[2]**

(c) When he was sampling with his net, Joshua also caught some other organisms.
He weighed all the organisms that he caught in one area.
His results are shown in the table.

Organism	Total mass in 1 m^2 in grams
freshwater shrimps	75
small fish	20
water beetles	35
water plants	300

(i) Use this data to draw the food chain that involves these four organisms.

[2]

(ii) Draw and label a pyramid of biomass for this food chain.

[3]

(Total 14 marks)

3 The diagram shows a section through a human heart during one part of its cycle.

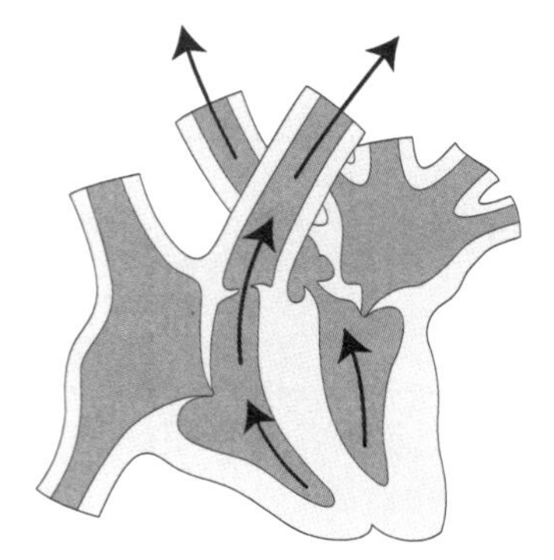

Leave blank

(a) Explain what is happening at the stage of the cycle that is shown in the diagram. One mark is awarded for a clearly ordered answer.

....

....

....

.... **[4+1]**

(b) It is important that the blood in the right side of the heart does not mix with the blood in the left side.
Explain why.

....

....

.... **[2]**

(Total 7 marks)

4 **(a)** A potted plant was kept in the dark for a day.
When one of the leaves was tested for starch with iodine solution it stayed brown.
Explain why this is.

....

....

.... **[3]**

(b) The plant was then set up as in the diagram.

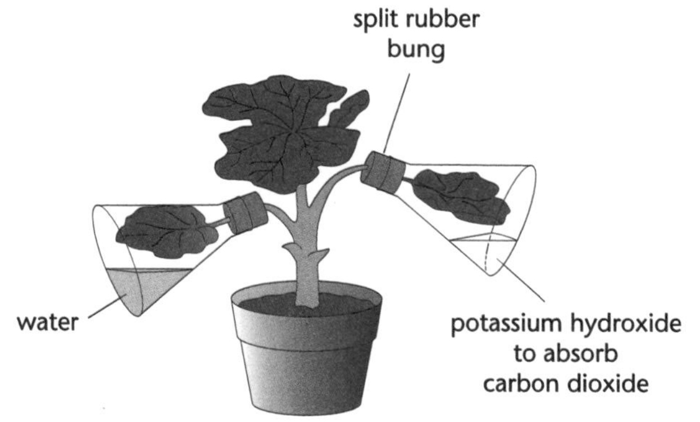

The plant with the flasks attached was then left in the light for several hours.
The leaves in the flasks were then tested for starch with iodine solution.

Leave blank

(i) The leaf enclosed in the flask with potassium hydroxide stayed brown.
The other leaf turned black.
Explain what this experiment proves.

....

....

.... **[2]**

(ii) Why was it necessary to enclose a leaf in a flask with water?

....

....

.... **[2]**

(Total 7 marks)

5 The menstrual cycle in the female is controlled by four hormones.
Two hormones LH and FSH are released from the pituitary gland.
Oestrogen and progesterone are released from the ovaries.

pituitary gland
FSH
inhibits
LH
the ovary before ovulation
stimulates the follicle
stimulates the corpus luteum
the ovary after ovulation
oestrogen
progesterone

(a) Both oestrogen and progesterone inhibit the release of FSH.
Write down one other effect that each has on the female body.

Oestrogen ....

....

Progesterone ....

.... **[2]**

Leave blank

(b) In order to prevent a woman becoming pregnant she may take the contraceptive pill.
This contains oestrogen and progesterone.
Use the information in the diagram to explain how the pill works.

....

....

.... **[3]**

(c) Some women have problems conceiving.
They may take a drug that blocks the action of oestrogen.

(i) This may lead to the woman carrying seven or eight embryos at once.
Explain why.

....

....

.... **[3]**

(ii) Doctors can operate in order to remove some of these embryos.
What are the possible arguments for and against this?

....

....

....

.... **[3]**

(Total 11 marks)

6 Clover is common plant that grows over most of Europe.

Clover is often eaten by slugs and snails.
The slugs and snails like to live in warm moist areas.
Some time ago a change in the DNA of a clover plant occurred.
This enabled the plant to produce a poison called cyanide in its leaves.

[turn over

Leave blank

(a) Write down the name given to the process that causes a change in the DNA of the clover.

.... **[1]**

(b) In many areas the cyanide-producing clover increased in numbers and became more common than the usual type.
Explain why this is.

....

....

.... **[3]**

(c) Scientists investigated the distribution of the cyanide-producing clover. They found that it was common in southern Europe but rare in northern Europe.
Suggest why this is.

....

....

.... **[3]**

(d) The production of cyanide is controlled by a dominant allele (Y).
The recessive allele (y) does not allow cyanide to be produced.

(i) Write down the genotype of a clover plant that is heterozygous for the cyanide-producing gene.

.... **[1]**

(ii) What proportion of cyanide-producing plants would be produced if one of these heterozygous plants fertilised another.
Use a genetic diagram to help you.

Proportion .... **[3]**

(Total 11 marks)

Leave blank

7 Mirlind designs an experiment to measure how fast a plant takes up mineral ions.
He grows a plant in a solution of mineral salts.
The tube allows him to bubble different gases through the solution.

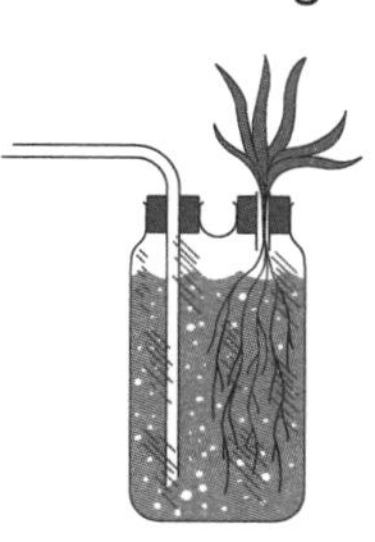

Mirlind then leaves the plant to grow for several days.

(a) However, before he could start the experiment, he noticed that the solution and the insides of the jar were turning green.

(i) Suggest why this happened.

....

.... **[2]**

(ii) Mirlind then placed black paper around the sides of the jar.
Explain why he did this.

....

.... **[2]**

(b) Mirlind could then start the experiment.
He measures the uptake of magnesium ions from the solution over a period of time.
He does this whilst bubbling oxygen through the solution.
His results are shown in the table.

Time in minutes	0	20	40	60	80	100
Amount of magnesium taken up in microgrammes	0.0	6.5	12	15.5	17.0	17.5

(i) Plot these results on the grid overleaf. **[4]**

[turn over

Leave blank

(ii) Finish the graph by drawing the best curve. [1]

(iii) Describe how the amount of minerals taken up by the plant varied during the 100 minutes.

....

....

.... [3]

(iv) Write down **one** use that the magnesium has in the plant.

.... [1]

(c) Mirlind's friend Rosie thinks that the plant takes up magnesium ions by active transport.
Mirlind tries to prove this by repeating the experiment using nitrogen gas rather than oxygen.
He finds that the plant takes up less magnesium.
Explain how this shows that Rosie was correct.

....

....

....

.... [4+1]

(Total 18 marks)

Centre number
Candidate number
Surname and initials

Letts Examining Group

General Certificate of Secondary Education

Science: Chemistry Paper

Higher tier

Time: 1 Hour

For Examiner's use only	
1	
2	
3	
4	
5	
6	
Total	

Instructions to candidates

Write your name, centre number and candidate number in the boxes at the top of this page.

Answer ALL questions in the spaces provided on the question paper.

Show all stages in any calculations and state the units. You may use a calculator.

Include diagrams in your answers where this may be helpful.

Information for candidates

The number of marks available is given in brackets **[2]** at the end of each question or part question.

The marks allocated and the spaces provided for your answers are a good indication of the length of answer required.

1 There have been many different Periodic Tables produced.
The diagram shows one produced by William Crookes. It puts the elements in order of increasing atomic mass but in spiral.

Leave blank

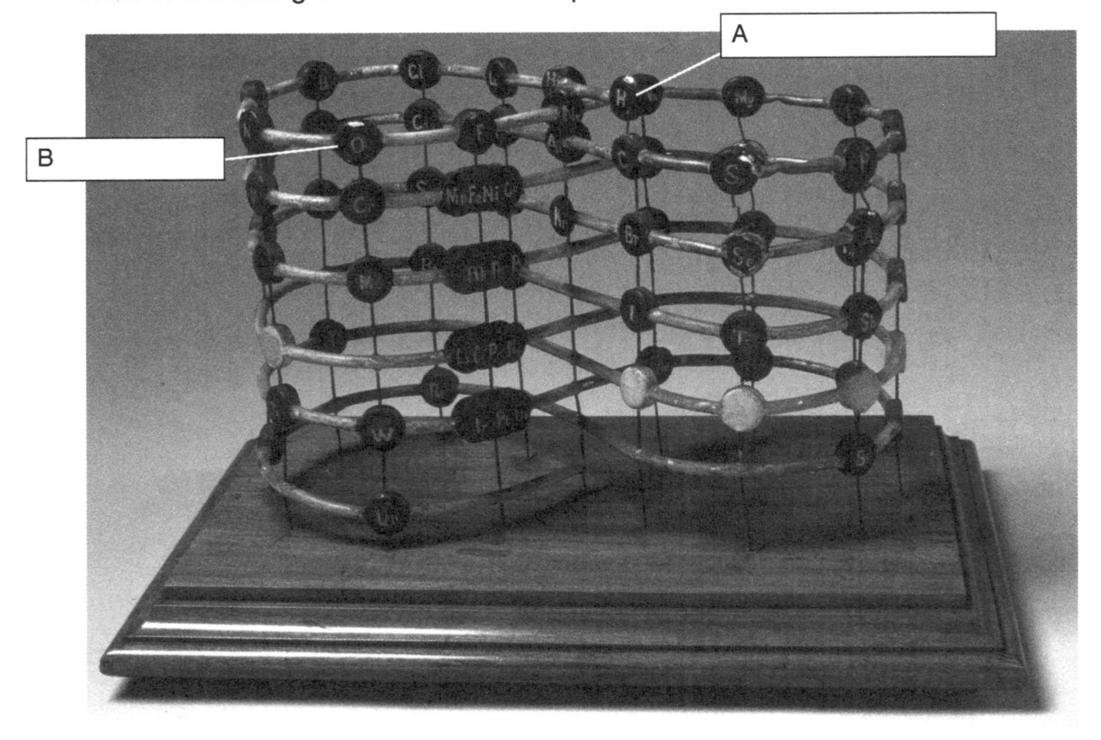

(a) Fill in the boxes A and B with the names of the two elements. **[2]**

(b) The elements phosphorus (P), arsenic (As), antimony (Sb) and bismuth (Bi) are in a vertical line in the spiral.
They have similar properties.
Look at the modern Periodic Table. There is an element in the same group that is not placed above phosphorus in this spiral.
Write down the name of this element. **[1]**

....

(c) Phosphorus is in group 5 of the Periodic Table.

(i) How many electrons are there in the outer shell of a phosphorus atom?

.... **[1]**

(ii) A phosphorus atom can form a phosphide ion.
Describe how this is done and give the charge on the ion. **[2]**

....

....

(iii) Phosphorus forms two oxides. Write the formulae of these two oxides.

.... **[1]**

(Total 7 marks)

2 Sphalerite is an ore of zinc. The flow diagram outlines how zinc can be extracted from sphalerite.

Leave blank

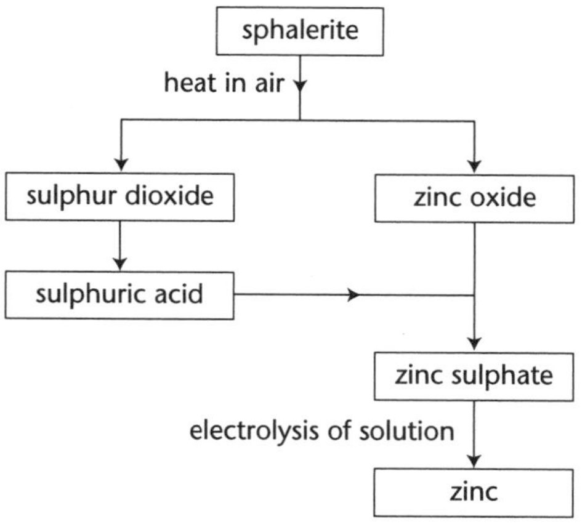

(a) Apart from zinc, name one other element present in sphalerite.

.... **[1]**

(b) Explain why sulphur dioxide should not be allowed to escape into the atmosphere for environmental reasons.

....

.... **[2]**

(c) Calculate the mass of zinc that can be extracted from 8.1 tonnes of zinc oxide. (Relative atomic mass: Zn = 65, O = 16)

Mass = ... tonnes **[3]**

(d) Write a balanced equation for the reaction of zinc oxide with dilute sulphuric acid.

.... **[3]**

(e) Explain how zinc can be extracted from zinc sulphate by electrolysis. Include an ionic equation in your answer. One mark awarded for a clearly ordered answer.

....

....

....

.... **[4 + 1]**

(Total 14 marks)

[turn over

Leave blank

3 An excess of marble chips (calcium carbonate) was added to 10 cm^3 of hydrochloric acid (40 g/dm^3). The volume of gas evolved was measured every 30 seconds.
The results are shown in the table.

Time in s	0	30	60	90	120	150	180	210
Volume of gas in cm^3	0	30	52	78	80	88	91	91

(a) Plot a graph of these results. Put time on the x-axis. Draw the best line of fit through the points. **[3]**

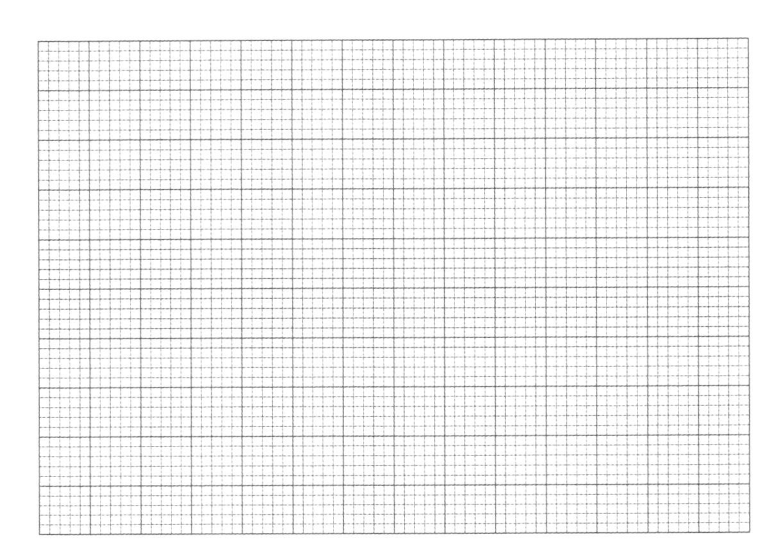

(b) Explain the shape of the graph.

....

.... **[2]**

(c) Write down **two** ways of making this reaction faster.

1 ....

2 .... **[2]**

(d) There is hardly any reaction if dilute sulphuric acid is used in place of hydrochloric acid. Suggest why.

....

.... **[2]**

(Total 9 marks)

Leave blank

4 The table gives information about some of the halogen elements in group 7 of the Periodic Table.

Element	Atomic number	Melting point in°C	Boiling point in°C	Atomic radius nm	Colour	Electron arrangement
fluorine	9	–220	–188	0.071	colourless	2,7
chlorine	17	–101	–34	0.099	greenish yellow	2,8,7
bromine	35	–7	58	0.114	orange	2,8,18,7
iodine	53	114	183	0.133	black	2,8,18,18,7

(a) Which halogen is liquid at room temperature and atmospheric pressure?

.... **[1]**

(b) How does the atomic radius change down the group?

.... **[1]**

(c) Explain how it can be shown that chlorine is more reactive than bromine using displacement reactions. Include an equation in your answer.

....

....

....

....

....

.... **[6]**

(d) Describe the bonding in magnesium fluoride. Include a 'dot and cross' diagram in your answer.

....

....

....

.... **[6]**

(Total 14 marks)

[turn over

Leave blank

5 **(a)** Write down the name of the series of hydrocarbons with the general formula C_nH_{2n+2}. .... **[1]**

(b) Draw the structure of the hydrocarbon with the formula C_3H_8. **[2]**

(c) Finish the balanced equations for the combustion of C_3H_8 in plentiful and limited supplies of air.

Plentiful supply of air

C_3H_8 + O_2 $\longrightarrow$ ....

Limited supply of air

C_3H_8 + O_2 $\longrightarrow$ .... **[4]**

(d) Another series of hydrocarbons has the general formula C_nH_{2n}. Ethene is the first member of this series.

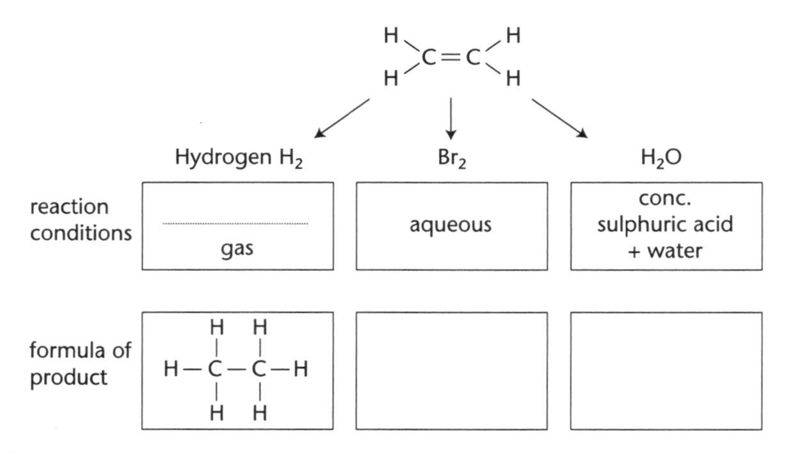

Complete the diagram. **[4]**

(Total 11 marks)

6 A sample of fertiliser contained ammonium sulphate, $(NH_4)_2SO_4$, as the only nitrogen fertiliser. The fertiliser was found to contain 14% nitrogen.

(a) Calculate the percentage of nitrogen in pure ammonium sulphate.
(Relative atomic masses: N = 14, H = 1, S = 32, O = 16)

Percentage = % **[3]**

(b) Calculate the percentage of ammonium sulphate in the fertiliser.

Percentage = % **[2]**

(Total 5 marks)

Centre number
Candidate number
Surname and initials

Examining Group

General Certificate of Secondary Education

Science: Physics Paper

Higher tier

For Examiner's use only	
1	
2	
3	
4	
5	
Total	

Time: 1 Hour

Instructions to candidates

Write your name, centre number and candidate number in the boxes at the top of this page.

Answer ALL questions in the spaces provided on the question paper.

Show all stages in any calculations and state the units. You may use a calculator.

Include diagrams in your answers where this may be helpful.

Information for candidates

The number of marks available is given in brackets **[2]** at the end of each question or part question.

The marks allocated and the spaces provided for your answers are a good indication of the length of answer required.

Leave blank

1 **(a)** When the voltage across a filament lamp is changed, the current also changes. Three circuits are shown in the diagram.

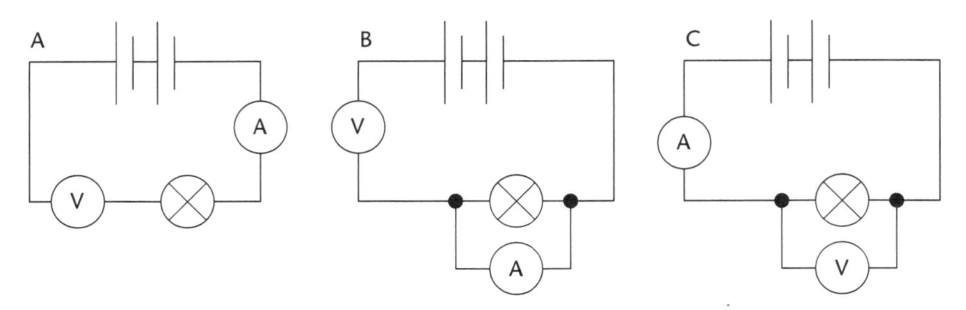

(i) Which circuit allows values of the current and voltage to be measured?

.... **[1]**

(ii) Which additional component would allow the current in the circuit to be varied?

.... **[1]**

(iii) State **one** other way of changing the current in the circuit.

.... **[1]**

(b) Here are three graphs that show how the current in a component changes when the voltage is changed.

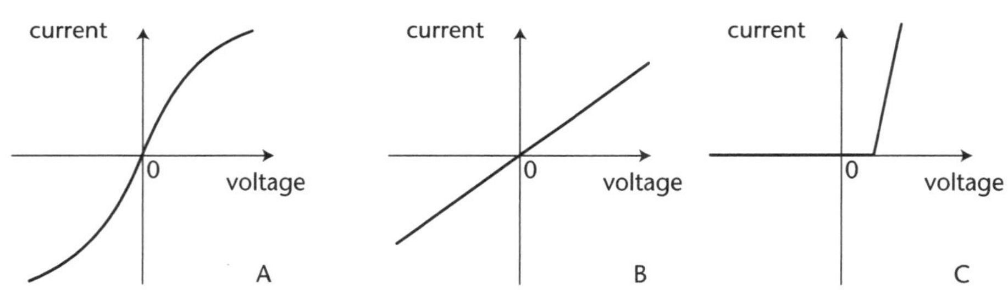

(i) Which graph represents a filament lamp?

.... **[1]**

(ii) Which graph represents a fixed resistor at a constant temperature?

.... **[1]**

(iii) Which graph represents a component that only allows current to pass in one direction?

.... **[1]**

(iv) Which graph represents a diode?

.... **[1]**

Leave blank

(c) The current in a lamp filament is 2.5 A when the voltage across it is 12 V. Calculate the power of the lamp.

....

....

.... **[3]**

(Total 10 marks)

2 A car accelerates from 20 m/s to 35 m/s in 6.0 s.

(a) Calculate the acceleration of the car.

....

....

.... **[3]**

(b) The car and driver have a total mass of 850 kg. Calculate the force needed to accelerate the car.

....

....

.... **[3]**

(c) Calculate the total kinetic energy when the car is travelling at 35 m/s.

....

....

.... **[3]**

(d) The car brakes. The brakes remove energy at the rate of 75 000 J/s. Calculate the time it takes for the brakes to stop the car.

....

.... **[2]**

(e) How far does the car travel during braking?

....

....

.... **[3]**

[turn over

Leave blank

(f) When the car is fully laden its mass is 1150 kg.
Explain how this affects the acceleration and the braking of the car.

....

....

.... **[3]**

(Total 17 marks)

3 The diagram represents a wave travelling along a rope.

(a) Which arrow could represent the vibrations of the rope? Explain how you can tell.

....

.... **[2]**

(b) Which arrow shows a distance equal to half a wavelength of the wave?

.... **[1]**

(c) Which arrow shows a distance equal to the amplitude of the wave?

.... **[1]**

(d) Explain why the rope cannot be used to model what happens when a sound wave is transmitted through the air.

....

.... **[2]**

(e) A wave of frequency 2.4 Hz passes along the rope. The wavelength of the wave is 0.35 m. Calculate the speed of the wave along the rope.

....

....

.... **[4]**

(Total 10 marks)

Leave blank

4 The generator in a power station consists of an electromagnet that rotates inside three sets of copper conductors.

(a) Explain why a current passes in the conductors when the electromagnet rotates.

....

.... **[2]**

(b) The current generated in each set of conductors is 6800 A at a voltage of 25 000 V.
Calculate the power generated in each set of conductors.

....

....

.... **[3]**

(c) The electricity passes from the generator to a transformer where the voltage is increased to 400 000 V before it passes into the national grid.

(i) The primary coil of the transformer has 1000 turns. Calculate the number of turns on the secondary coil.

....

....

.... **[3]**

(ii) Explain why the voltage is increased before the electricity is transmitted.

....

....

.... **[3]**

(d) A coal-burning power station has an efficiency of 40%.
Describe what happens to the energy released from the burning coal. Plus 1 mark for correct spelling, punctuation and grammar.

....

....

.... **[3+1]**

(Total 14 marks)

[turn over

Leave blank

5 The diagram shows a satellite in orbit around the Earth.

(a) Add an arrow to the diagram to show the force on the satellite. **[1]**

(b) Which word from the list describes the type of force acting on the satellite? Underline your choice. **[1]**

electric **gravitational** **magnetic** **nuclear**

(c) The time it takes a satellite to complete one orbit of the Earth depends on its height above the Earth's surface.
Some orbit times are given in the table.

Height above Earth's surface in millions of metres	Orbit time in hours
0	1.5
12	7.0
26	16.0
40	27.5
50	37.0

(i) Use the grid to draw a graph of height above the Earth's surface against orbit time.

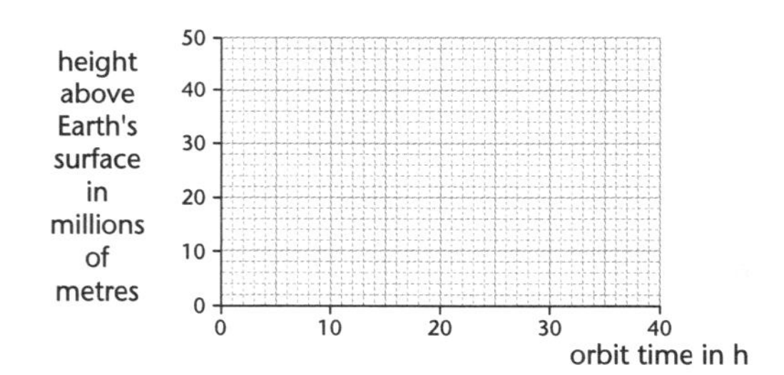

[3]

A satellite orbits the Earth at a height of 36 million metres.

(ii) Use the graph to find the orbit time of the satellite.

.... **[1]**

(iii) Explain why this satellite stays above the same point on the Earth's surface.

....

.... **[2]**

(iv) Suggest a use for this satellite.

.... **[1]**

(Total 9 marks)

Answers to mock examination Papers

Biology

1 (a)(i) Optic nerve labelled A [1]
(ii) Retina labelled B [1]
(iii) Iris labelled C [1]
(b) The rays of light are focused behind the eye [1]
Do not focus on the retina [1]
(focused behind the retina = 2)
(c) Rods cannot detect colour, cones can
Rods can work in dimmer light than cones
Rods detect less detail about the image
any two [2]

Examiner's tip

Remember that the fovea contains cones only so that when we look directly at an object we are using cones to see the detail.

2 (a) (i) $C_6H_{12}O_6$ [1]
$6H_2O$ [1]
(ii) Movement / Active transport / Sensitivity
any one [1]
(b) (i) To make it a fair test [1]
So that he could compare the different areas [1]
(ii) Shrimps like to live in fast flowing areas [1]
But avoid the fastest flowing sections [1]
(iii) The faster the flow of the water, the more oxygen it contains
So more available for respiration
In fastest areas they may get washed away/damaged any two [2]
(c)(i) water plants → freshwater shrimps → water beetles → small fish [2]
(ii)

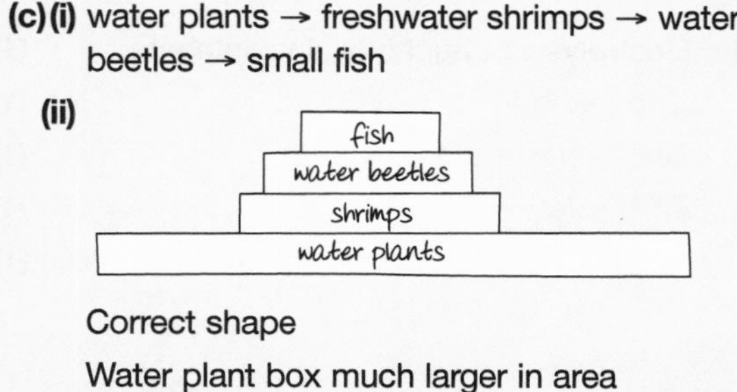

Correct shape [1]
Water plant box much larger in area [1]
Correctly labelled [1]

Examiner's tip

A common Higher tier question is to ask why this is a pyramid. Make sure that you can explain losses of biomass and energy along food chains.

3 (a) Ventricles are contracting
Blood is passing out of the arteries
Aorta and pulmonary arteries named
Atria relaxed
Filling up with blood
Tricuspid and bicuspid valves shut any four [4]
plus 1 mark for a clearly ordered answer [1]
(b) Left side contains oxygenated blood, right side is deoxygenated [1]
If they mix, deoxygenated blood would be pumped to the body [1]

4 (a) Starch absent [1]
Cannot photosynthesise and make starch in the dark [1]
Stores used up in the dark [1]
(b)(i) Carbon dioxide is needed for photosynthesis [1]
No carbon dioxide then no starch production [1]
(ii) As a control/comparison [1]
In order to make sure that the flask was not altering starch production [1]

5 (a) Oestrogen: repairs wall of uterus/female secondary sexual characteristics [1]
Progesterone: prevents wall of uterus breaking down [1]

Examiner's tip

Natural selection questions are always looking for the same pattern in the answer. Certain variation causes an advantage so that certain individuals can survive, reproduce and pass on the variation.

(b) Both hormones stop action of FSH [1]
Follicle not stimulated [1]
Ovulation does not occur [1]

(c) (i) FSH not inhibited [1]
More follicles stimulated [1]
Many eggs released at one time [1]

(iii) For: All babies may die if all left
May cause dangers to mother carrying many embryos
Allows some embryos to survive and be born
Against: Danger to mother of performing operation
Killing human life
Having to choose which survive and which die any three (at least one from each section) [3]

6 (a) Mutation [1]

(b) Less likely to be eaten by slugs and snails
Survive to reproduce
Produce offspring that can make cyanide/contain the allele
Natural selection any three [3]

(c) Slugs and snails like warm places
More of them in southern Europe
Large advantage here to make cyanide
In northern Europe it may be a disadvantage any three [3]

(d)(i) Yy [1]

(ii) Gametes Y or y × Y or y [1]
Genotypes YY Yy yY yy [1]
Proportion one quarter/one in four/one to three/25% [1]

7 (a)(i) Plants/algae in the water [1]
Were reproducing [1]

(ii) To kill the algae [1]
By stopping them photosynthesising [1]

(b)(i) Time on x-axis, mineral uptake on the y-axis [1]
Reasonable scales [1]
Correct plots [2]

(ii) Smooth best curve [1]

(iii) Taken up at a steady rate at first up to about 60 minutes
Uptake starts to slow down
Very little/no more taken up after about 80 minutes any three [3]

(iv) Used to make chlorophyll [1]

(c) Oxygen needed for respiration/less respiration with nitrogen [1]
Respiration releases energy/less respiration [1]
Active transport needs energy [1]
Because it transports ions against a concentration gradient [1]

Quality of written communication mark. This mark should only be awarded if the candidate attempts to address the question and the quality of the description makes the meaning clearer. [1]

Examiner's tip

This is a hard question. Active transport is definitely an A/A topic.*

Chemistry

1 (a) A – Hydrogen [1]
B – Oxygen [1]

(b) Nitrogen [1]

(c)(i) 5 [1]

(ii) Each phosphorus atom gains three electrons [1]
Ion has 3- charge [1]

(iii) P_2O_5 and P_2O_3 both required [1]

2 (a) Sulphur [1]

(b) Sulphur dioxide reacts with air and water to form sulphuric acid [1]
Causes acid rain [1]

(c) RFM of ZnO = 65 + 16 = 81 [1]
81 tonnes of ZnO produces 65 tonnes of Zn [1]
6.5 tonnes [1]

(d) $ZnO + H_2SO_4 \rightarrow ZnSO_4 + H_2O$ [3]
1 mark for reactants, 1 for product, 1 for balancing

Examiner's tip

The balancing mark here is easy as the equation is already balanced.

(e) Electrolysis of <u>aqueous</u> zinc sulphate [1]
Zinc cathode [1]
Zinc deposited on cathode [1]
$Zn^{2+} + 2e^- \rightarrow Zn$ [1]
plus 1 mark for a clearly ordered answer [1]

3 (a)

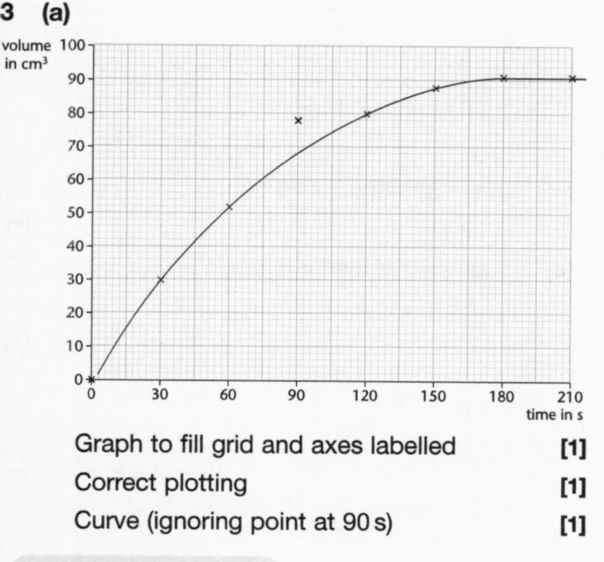

Graph to fill grid and axes labelled	**[1]**
Correct plotting	**[1]**
Curve (ignoring point at 90 s)	**[1]**

Examiner's tip

This is a typical high demand graph. You have to ignore the anomalous point.

(b) Graph becomes less steep as acid is used up **[1]**

Graph becomes level when all acid used up **[1]**

(c) Increasing concentration of acid

Increasing temperature

Crushing marble into smaller pieces

any two **[2]**

Examiner's tip

You must relate your answer to the situation given. Adding a catalyst is not acceptable as there is no suitable catalyst.

(d) Calcium sulphate is not very soluble **[1]**

Layer formed on marble chip stopping acid and marble coming into contact **[1]**

Examiner's tip

Part (d) is very hard and targeted at A.*

4 (a) Bromine **[1]**

(b) Increases **[1]**

(c) Add chlorine **[1]**

To potassium bromide solution **[1]**

Solution changes from colourless to red/brown **[1]**

$Cl_2 + 2KBr \rightarrow 2KCl + Br_2$

or $Cl_2 + 2Br^- \rightarrow 2Cl^- + Br_2$ **[3]**

1 mark for formulae reactants, 1 mark for products, for balancing

Examiner's tip

The common mistake here is not to give the colour of potassium bromide solution before the reaction.

(d) Ionic bonding

Each magnesium loses two electrons

Each fluorine atom gains one electron

One magnesium ion and two fluoride ions

Held together by electrostatic bonding

any four **[4]**

$[F]^- \; Mg^{2+} \; [F]^-$ or $[F]^- \; [Mg]^{2+} \; [F]^-$

1 mark for electrons correct, 1 mark for charges **[2]**

5 (a) Alkanes **[1]**

(b)

```
    H   H   H
    |   |   |
H — C — C — C — H
    |   |   |
    H   H   H
```

one mark awarded if there is a slight mistake **[2]**

(c) $C_3H_8 + 5O_2 \rightarrow 3CO_2 + 4H_2O$ **[2]**

$2C_3H_8 + 7O_2 \rightarrow 6CO + 8H_2O$ **[2]**

In each case there is one mark for the formulae of the products and one for balancing

(d) (Nickel) catalyst **[1]**

Heat **[1]**

```
    Br  Br
    |   |
H — C — C — H
    |   |
    H   H
```

[1]

```
    H   H
    |   |
H — C — C — OH
    |   |
    H   H
```

[1]

6 (a) RFM of ammonium sulphate

= 28 + 8 + 32 + 64 = 132 **[1]**

Percentage = $\frac{28 \times 100}{132}$ **[1]**

= 21.2% **[1]**

(b) 21.2% corresponds to 100% ammonium sulphate

14% corresponds to

$\frac{100}{21.2} \times 14\%$ ammonium sulphate **[1]**

= 66% **[1]**

Physics

1(a) (i) C [1]

(ii) A variable resistor [1]

(iii) Change the voltage [1]

Examiner's tip

Avoid answers such as 'use a bigger battery'. It is the battery voltage that is important, not the physical size.

(b (i) A [1]

Examiner's tip

The resistance of a lamp filament increases as the filament gets hotter; this is why the slope of the graph changes.

(ii) B

(iii) C

(iv) C [1]

Examiner's tip

The resistance of a fixed resistor at constant temperature stays the same so the current/voltage graph has a constant gradient.

(c) power = current × voltage [1]

= 2.5 A × 12 V [1]

= 30 W [1]

Examiner's tip

Remember, you must have the correct unit to gain the final mark.

2 (a) acceleration = increase in velocity ÷ time taken [1]

= 15 m/s ÷ 6.0 s [1]

= 2.5 m/s^2 [1]

Examiner's tip

Take care with the unit when working out accelerations. Most candidates at GCSE give the unit as m/s, losing one mark out of three.

(b) force = mass × acceleration [1]

= 850 kg × 2.5 m/s^2 [1]

= 2125 N [1]

(c) kinetic energy = $\frac{1}{2}mv^2$ [1]

= $\frac{1}{2}$ × 850 × (35)^2 [1]

= 520 625 J [1]

Examiner's tip

Common errors when calculating kinetic energy are forgetting to square the speed and forgetting the half.

(d) time = 520 625 J ÷ 75 000 J/s [1]

= 6.9 s [1]

(e) distance travelled = average speed × time [1]

= 17.5 m/s × 6.9 s [1]

= 121 m [1]

Examiner's tip

*If your answer to **(e)** is 242 m, you have not taken account of the fact that in braking from 35 m/s to rest (0 m/s), the average speed of the car is half that of the initial speed.*

(f) The acceleration is reduced [1]

The deceleration is reduced [1]

The braking or stopping distance is increased [1]

3 (a) D [1]

The wave shown is transverse [1]

(b) E [1]

Examiner's tip

B shows a distance equal to one quarter of a wavelength and A shows one wavelength. One wavelength is one complete cycle of a wave, a crest and a trough for a transverse wave or a compression and a rarefaction for a longitudinal wave.

(c) C [1]

Examiner's tip

A common error in this question is to answer D. The amplitude is the maximum displacement from the mean or rest position, not the difference between the extremes of displacement.

(d) Sound is a longitudinal wave [1]

A rope cannot easily demonstrate longitudinal waves. [1]

(e) speed = frequency × wavelength [1]

= 2.4 Hz × 0.35 m [1]

= 0.84 m/s [1]

Quality of written communication mark – calculation set out in a clear, logical order [1]

4 (a) The changing magnetic field [1]

Induces or causes a current [1]

Examiner's tip

Always emphasise the change in the magnetic field when explaining electromagnetic induction.

(b) power = current × voltage **[1]**

= 6800 A × 25 000 V **[1]**

= 170 000 000 W (1.7×10^8 W) **[1]**

(c)(i)

$$\frac{\text{primary turns}}{\text{secondary turns}} = \frac{\text{primary voltage}}{\text{secondary voltage}}$$ **[1]**

$$\text{secondary turns} = \frac{\text{primary turns} \times \text{secondary voltage}}{\text{secondary turns}}$$ **[1]**

$$\frac{1000 \times 400\,000}{25\,000} = 16\,000$$ **[1]**

Examiner's tip

It is usually easier to use ratios when calculating transformer turns or voltages. In this case the ratio of the voltages is 1:16, so the numbers of turns must be in the same ratio.

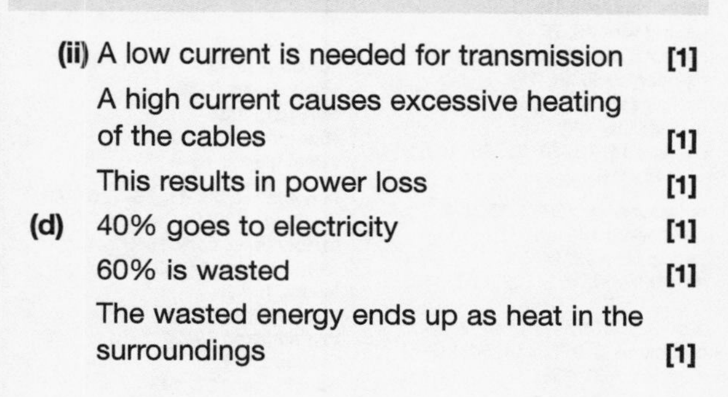

(ii) A low current is needed for transmission **[1]**

A high current causes excessive heating of the cables **[1]**

This results in power loss **[1]**

(d) 40% goes to electricity **[1]**

60% is wasted **[1]**

The wasted energy ends up as heat in the surroundings **[1]**

5 (a) The arrow should point towards the centre of the Earth **[1]**

(b) Gravitational **[1]**

Examiner's tip

Planets, moons and satellites are kept in orbit around a larger body because of the attractive gravitational forces between them. These forces act from centre to centre of the two objects. A common error is to draw the force arrow in the direction of motion.

(c) (i) This is the completed graph **[2]**

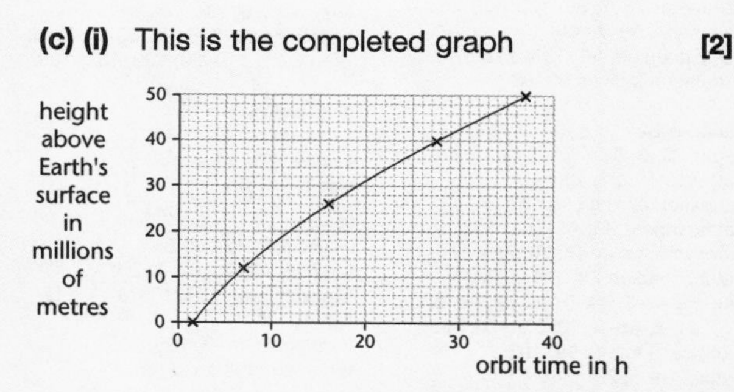

Correct plot (allow one error)

award one mark if there are two or three errors **[2]**

Curve drawn **[1]**

(ii) 24 hours **[1]**

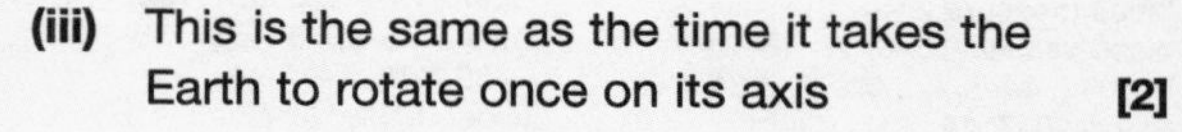

(iii) This is the same as the time it takes the Earth to rotate once on its axis **[2]**

award 1 mark for identifying this time with the length of a day

(iv) Communications/television/telephone **[1]**

Index